Design and Test Strategies for 2D/3D Integration
for NoC-based Multicore Architectures

Kanchan Manna • Jimson Mathew

Design and Test Strategies for 2D/3D Integration for NoC-based Multicore Architectures

 Springer

Kanchan Manna
Indian Institute of Technology Patna
Patna, Bihar, India

Jimson Mathew
Indian Institute of Technology Patna
Patna, Bihar, India

ISBN 978-3-030-31309-8   ISBN 978-3-030-31310-4   (eBook)
https://doi.org/10.1007/978-3-030-31310-4

This Springer imprint is published by the registered company Springer Nature Switzerland AG.
The registered company address is: Gewerbestrasse 11, 6330 Cham, Switzerland

Preface

Research interest into the Network-on-Chip (NoC) concept has increased considerably over the last few years in both academia and industry. Academic examples of complete systems include the Nostrum, Mango, SOCBUS and Xpipes while in industry Spidergon, Aetherial, Silistix and Arteris are commercially available. To improve performance of the systems (which are synthesized on Silicon), on-chip device density and diversity have been increasing. For such on-chip integration or Very Large Scale Integration (VLSI), the traditional System-on-Chip (SoC) based approach is not suitable due to its several limitations: scalability, bandwidth and latency. The technique does not scale well with increasing number of devices. Bandwidth limitation of bus has led the SoC designer to look for better communication technique. In addition, technology scaling causes severe synchronization problems in global interconnect, unpredictable delay and high power consumption. As an alternative, NoC has evolved as a standard to mitigate such problems.

In NoC, devices or IP-cores communicate with each other using a network. The network consists of routers and links. Routers are connected via links using a topology. The performance of traditional or two-dimensional (2D) NoC-based systems depends on the diameter of the network topology, as it determines the maximum delay in such systems. In addition, demand of interconnection scaling and device diversity have suggested an alternative three-dimensional (3D) device integration technique. The 3D integration technology offers the promise of being a new way of increasing the system performance without scaling. The promises are due to several characteristics of 3D integration: decreased wire length and thus reduced interconnect delay, increased number of interconnects between modules, and the ability to allow various materials, process technologies and functions in a single chip. In this approach, several silicon dies are stacked and connected by using the vertical interconnections. The length, delay and power consumption of vertical and horizontal interconnections are typically asymmetric. Vertical interconnections outperform horizontal ones. For this inherent property, 3D integration can be a good choice for designing a complex system. To design three-dimensional NoC (3D NoC) based systems, the primary components of the NoC, that is, routers, should have additional two ports to connect two additional neighbours in vertical direction.

Through-Silicon-Via (TSV) has evolved as a viable communication channel across vertical direction of an extended router, as TSV provides low latency and low power consumption. However, the TSV overhead including area, manufacturing cost, routing congestion and yield loss can increase significantly as the number of TSVs increases in a system. Moreover, it is expected that the maximum number of TSVs in a high-performance 3D chip will be increased in every year. The dimension of the TSV is key to 3D circuit designers, since it directly impacts exclusion zones, where designers cannot place the transistors. Therefore, TSVs are expected to be limited in number. They should be spread out for reducing the routing congestion in the 3D NoC-based systems. The stringent constraints over number of TSV usage in a 3D system as well as their position can make 3D NoC routers heterogeneous. That is, some routers in a layer can have the TSV, whereas others do not. A minimum distance has to be maintained between adjacent 3D-routers to take care of the TSV geometry. This kind of router configuration can increase the average distance among the IP-cores as compared to a traditional or fully connected 3D NoC. So such kind of configuration can affect the overall system performance. The challenge of combining both the approaches of 3D and NoC is to come up with the association of cores of the fabric to the tasks of the application and proper placement of a limited number of TSVs, making efficient use of the available hardware resources and satisfying the communication needs of all the tasks.

Power and temperature are two roadblocks for designing the sophisticated systems. It is also predicted that power density in the systems will increase in successive generation of scaling technology. High-performance circuits consume a significant amount of power, due to diverse functionalities and high frequency of operation. The consumed power dissipates as heat. The high operating temperature of the systems may lead to its unreliable operations. The increased power density can also lead to an increase in several parameters: data activity, leakage power dissipation and electro-migration. The high temperature can create vicious thermal cycle acting as the positive feedback to increase leakage power further. Improper thermal gradient, in turn, increases the failure rate of the chip. Interconnect delay can also increase due to increase of temperature. In addition, temperature increase is a global phenomenon as temperature of a module is increased by the diffusion of temperature from the surrounding modules. In addition, the device packaging and cooling technologies imposed a limit on the peak temperature for such systems. Therefore, it is very challenging to deal with thermal heating problem in today's system design techniques.

Thermal heating problems are also exacerbated with the transition from a 2D-based system to 3D-based systems. The 3D integration can provide more device density, bandwidth and speed. On the other hand, the amount of heat per unit footprint increases with the integration of more devices, resulting in high temperature. In addition, the heat sink is located far away from some of the layer. Therefore, it is very challenging to remove heat from the 3D-based systems.

The device scaling technology has also increased the susceptibility to internal defects in such systems. So, manufacturing tests of such systems are crucial, and this is a complex and time-consuming process. Due to stress on time-to-market, test engineers focus on the reduction of testtime and perform parallel tests of modules. Due to aggressive technology scaling into the nanometre regime, power consumption is also becoming a significant burden as system designers set a limit on maximum power consumption and peak temperature. The limits are imposed by device packaging and cooling technology on the peak temperature. Moreover, power consumption during manufacturing tests is more ($2\times$) as compared to normal operation. In addition, peak power consumption is often significantly higher ($30\times$) than the average power values. The consumed power leads to high temperature and creates hotspots, which in turn leads to failure of good parts, resulting in yield loss. Thermal safety during testing of such systems is a very challenging problem, including 3D-based systems due to stacking of layers. This book highlights the research activities of the aforementioned challenges in the domain of NoC-based system design and test. This book is suitable for the scholar and the designer at industry who are working on such domain.

Organization This book is organized into nine chapters. A summary of the chapters is presented in the following paragraphs.

Chapter 1 presents the evolution of NoC-based systems, with different issues: effect of three-dimensional (3D) integration technology, power and temperature, and testing of such systems. It also highlights the shortcomings of SoC-based system design technique. It also presents the importance of PSO technique as compared to other evolutionary approaches.

Chapter 2 describes several alternative solutions for the above-mentioned problems and also discusses the shortcomings of the proposed approaches. It also discusses about several design-time and runtime solutions for the problems.

Chapter 3 highlights an iterative heuristic technique to place the limited number of TSVs in consultation with the application, in 3D NoC-based systems. This approach is an extension of Kernighan–Lin (KL) heuristic technique. We have augmented the basic KL approach by incorporating the iterative improvement phase. To improve the solution quality, this phase has applied different operations: flipping and swapping. The flow of the proposed technique and experimental results are also presented here.

Chapter 4 describes a constructive heuristic technique to place the limited number of TSVs by consulting with the application, in 3D NoC-based systems. The motivation of such heuristic over iterative heuristic is presented in this chapter. Constructive heuristic technique explores the possible solution more as compared to the iterative heuristic technique. This chapter presents the overall workflows of the proposed approach, experimental results, including comparison of iterative-based approach.

Chapter 5 presents the placement of the limited number of TSVs in consultation with the application, in 3D NoC-based systems by using Integer Liner Programming (ILP). It also highlights the shortcomings of ILP, while considering the bigger benchmarks. As a remedy, PSO, a meta-heuristic, technique is also proposed. In addition, the basic PSO technique is augmented using several innovative operations, to improve the solution quality. This chapter presents the overall working principle of two proposed approaches, experimental results, including comparison of iterative-based and constructive-based approaches.

Chapter 6 presents a thermal model to compute the temperature of the system. It formulates the thermal-aware application mapping technique to improve the thermal safety of 2D NoC-based systems. First, the problem is formulated by using ILP. It also highlights the shortcomings of ILP, while dealing with the bigger benchmarks. As a remedy to such problem, a discrete PSO, a meta-heuristic, technique is also proposed. In addition, the basic PSO technique is augmented using several innovative operations, to improve the solution quality. This chapter presents the overall workflows of two proposed approaches, experimental results, including comparison of other state-of-the-art approaches. A trade-off between performance and thermal safety is also described in this chapter.

Chapter 7 describes the thermal-aware application mapping technique to improve the thermal safety of 3D NoC-based systems. It presents two approaches to solve the heating problem in 3D-based system. In one approach, some tolerance has been imposed onto the performance (communication cost) to maintain the thermal safety, whereas, in other approach, some extra silicon area has reserved to place the thermal vias. A PSO, a meta-heuristic, technique is used in these approaches. In addition, to improve the solution quality, the basic PSO technique is augmented using several innovative operations. This chapter presents the overall working principle of two proposed approaches, experimental results, including comparison between the proposed approaches.

Chapter 8 highlights the temperature issues during the test of such systems, including the improvement of testtime. The testtime of such systems is improved by judicious testing of non-preemptive and preemptive modules or IP-cores. The problem is formulated using an ILP. The shortcoming of ILP-based approach is also highlighted here. As a remedy, a PSO-based, a meta-heuristic, technique is used. Next, testing problem is extended to temperature aware, for 2D as well as 3D environments and PSO-based technique is adopted for this purpose. In addition, the basic PSO technique is augmented using several innovative operations, to improve the solution quality. This chapter presents the overall workflows of the proposed approaches, experimental results, including comparison of other state-of-the-art approaches. A trade-off between testtime and temperature is also described in this chapter.

Chapter 9 explores several future research directions, such as application-specific systems design, integrated mapping and scheduling, reliability and fault tolerant technique and multi-cast-based NoC testing.

Patna, Bihar, India   Kanchan Manna
Patna, Bihar, India   Jimson Mathew

Contents

Chapter 1
Introduction

In sophisticated embedded VLSI products, a single chip implementation integrating several Intellectual Property (IP) cores for performing various functions and possibly operating at different clock rates is quite common. This implementation is traditionally known as System-on-Chip (SoC). The SoC-based system design methodology focuses on the computational aspects of the problem. However, the number of components in a single chip and their performances continue to increase. To address complex real-life applications, it is required to have multiple processors which can cohesively communicate and provide high parallelism. This, in turn, has resulted in Chip Multi-Processing (CMP) systems to provide scalable computational power. Hundreds of processing cores are integrated on the SoC platform to build Multi-Processor System-on-Chip (MPSoC) in deep submicron (DSM) technology. In these systems, the design of communication architecture plays a major role in defining the area, performance and energy consumption of the overall system.

1 System-on-Chip to Network-on-Chip: A Paradigm Shift

Traditionally, the shared medium arbitrated bus has been used as a communication architecture in SoC-based systems. However, it has several limitations, when several IPs are connected through a bus, such as scalability, bandwidth and latency (Dally and Towles 2001). The architecture does not scale with the number of cores attached (Grecu et al. 2004). Bandwidth limitation of buses has led the SoC designers to look for better communication alternatives. Furthermore, technology scaling causes severe synchronization problems in the global interconnect, unpredictable delay and high power consumption. Satisfactory performance can be expected only when a chip contains up to ten cores. However, as in the many-core era, the number of cores residing on an SoC increase continuously; the focus of optimization is shifting from computation to communication. Point-to-Point (P2P) communication architecture

© Springer Nature Switzerland AG 2020
K. Manna, J. Mathew, *Design and Test Strategies for 2D/3D Integration for
NoC-based Multicore Architectures*, https://doi.org/10.1007/978-3-030-31310-4_1

can be a good choice to mitigate the problem of global buses. However, the number of links used in this architecture grows exponentially with the cores residing to the system. For a large system, it may create routing problem (Bjerregaard and Mahadevan 2006). A centralized crossbar switch mitigates some of the limitations of the bus. However, connecting a large number of cores to a single switch is not very effective, as it is not ultimately scalable, and thus, is an intermediate solution only (Bjerregaard and Mahadevan 2006). In the many-core regime, individual processor speed has improved significantly over the technology generations. As a result, communication architecture has become the roadblock, limiting the overall system performance. Several research groups from academia and industry have started to find out suitable communication architectures for next generation many-core based SoCs. In the process, Network-on-Chip (NoC) has evolved as a standard to design the advanced Multi-Processor Systems-on-Chip (MPSoCs). It provides better predictability, lower power consumption, greater scalability and fault-tolerance compared to the previously known solutions for on-chip communication (Ho et al. 2001; Dally and Towles 2001; Benini and Micheli 2002). In an NoC, the processing components (known as IP-cores) communicate between themselves using an under-lying fabric of routers, connected in some topological fashion. Individual cores are attached to routers through network interface (NI) module (Dally and Towles 2001; Benini and Micheli 2002). The traditional data signal exchange between IP-cores gets replaced by message passing through router network (Dally and Towles 2001; Benini and Micheli 2002).

An NoC structure consisting of heterogeneous IP-cores (CPU, DSP, etc.) has been presented in Fig. 1.1. The IP-cores communicate with each other via the network and the network interface (NI) modules. The NI serves as a gateway

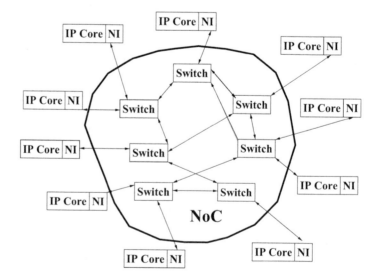

Fig. 1.1 An NoC-based systems

to convert computation to communication and vice versa. The network consists of routers and communication links between them. Length of the communication channel is primarily determined by the area occupied by the IP-cores, which is typically unaffected by the network structure. Due to the advancement in CMOS scaling technology, a large number of IP-cores can be incorporated into a single die. However, the network diameter of an NoC can increase with more number of cores added into a single die, which in turn increases the communication delay and power consumption. Hence, the overall system performance decreases with the increase in diameter.

2   NoC-Based Multi-Core Systems with Three-Dimensional (3D) Integration Technology

For the demand of interconnection scaling, three-dimensional (3D) integration has proposed as a solution. The 3D integration technology offers the promise of being a new way of increasing the system performance without scaling. The promise is due to 3D integration's several of characteristic features: decreased wire length and thus reduced interconnect delay, increased number of interconnects between modules, and the ability to allow various materials, process technologies and functions in a single chip. In this approach, several silicon-dies are stacked and connected by using the vertical interconnections. The length, delay and power consumption of vertical and horizontal interconnections are typically asymmetric (Xu et al. 2011). Vertical interconnections outperform horizontal ones. For this inherent property, 3D integration can be a good choice for designing a complex system. To design three-dimensional NoC (3D NoC) based systems, the primary components of the NoC, that is, routers should have the facility to support 3D layout. The easiest way to extend a 2D router to 3D can be to add two more ports to connect two additional neighbours in vertical direction with proper extension of associated components: crossbar. The communication channel across vertical direction of an extended router can be implemented by several techniques like wire bonding, micro-bump, contact-less interconnection and Through-Silicon-Via (TSV) (Davis et al. 2005). Among them, TSV is the most viable solution due to its low latency and low power consumption (Xu et al. 2011). However, the TSV overhead including area, manufacturing cost, routing congestion and yield loss, can increase significantly as the number of TSVs increases in a system (Pasricha 2012; Xu et al. 2011). Moreover, as per the International Technology Roadmap for Semiconductors (ITRS), the maximum number of TSVs in a high-performance 3D chip is about 1000 in 2012 and expected to increase further by 1000 in every year (Semiconductor Industry Association 2007). The dimension of the TSV is key to 3D circuit designers, since it directly impacts exclusion zones, where designers cannot place the transistors. Therefore, TSVs are expected to be limited in number. They should be spread out for reducing the routing congestion in the 3D NoC systems.

The stringent constraints over number of TSV usage in a 3D system as well as their position can make 3D NoC-routers heterogenous. That is, some routers in a layer can have the TSV, whereas others do not. A minimum distance has to be maintained between adjacent 3D routers to take care of the TSV geometry. This kind of router configuration can increase the average distance among the cores and reduce the bisection width as compared to a fully connected 3D NoC. As the bisection width and the average distance correspond to the number of parallel communication and average latency in NoCs, degradation of these parameters have negative effect on the overall system performance. The challenge of combining both the approaches of 3D and NoC is to come up with the association of cores of the fabric to the tasks and proper placement of limited number of TSVs, making efficient usage of the available hardware resources and satisfying the communication needs of all the tasks.

3 Power and Temperature Issues in NoC-Based Multi-Core Systems

In NoC-based multi-core systems, power and temperature have become two dominant constraints (Shang et al. 2006). Power density in the processing units has increased in the recent past. It is expected to increase further in successive the generations due to scaling in the features of CMOS technology, other than the reduction in operating voltage (Semiconductor Industry Association 2007). High-performance circuits consume a significant amount of power due to their variety of functionalities and higher frequency of operation. A significant portion of consumed power gets directly dissipated as heat. The high operating temperature of the systems may lead to its unreliable operation. The increase in power density can also lead to an increase in several parameters: data activity, leakage power dissipation (exponential in nature) and electro-migration, resulting in even higher temperature. The high temperature can create vicious thermal cycle acting as the positive feedback to increase leakage power further. Improper thermal gradient, in turn, increases the failure rate of the chip. Thermal hotspots can get created due to uneven power distribution which in turn decreases the system performance and lifetime (mean time to failure). Interconnect delay can also increase due to an increase in temperature (about 5% for every 10 °C rise in temperature) (Quaye 2005). The high instantaneous temperature in the IC can lead to catastrophic failures, as well as long-term degradation in the IC and packaging materials. These may eventually lead to system failures (FlipChip 2005). Therefore, thermal heating plays a major role in today's IC design. It is required to ensure the good thermal behaviour of such a system in design time, even when it dissipates the maximum power. To cope with this problem, one solution is to design efficient heat sink, so that internally generated heat can transfer quickly to the ambience or some other cooling techniques. Various cooling techniques are available in the literature. However, it is predicted by ITRS that the power density of 14 nm technology node will be greater than $100 \, W/cm^2$ and the thermal resistance between junction to ambient will be less than $0.2 \, °C$ (Semiconductor Industry Association 2009). It is very important to keep the thermal resistance limited to moderate value, as this may increase the packaging and overall product cost.

One way to address the high and uneven distribution of temperature across the system chip is to make sure that the IP-blocks are placed in such a way that they can even out the temperature profile of the system. This necessitates the placement of IP-blocks to be guided not only by their communication requirements but also their temperature profile. However, such a solution is suitable for dedicated systems with specific applications only. In the generic NoC-based system, such a problem can be mitigated by using a proper association between tasks in the application and cores in the topology, honoring the communication requirement of the application.

Thermal problems are also exacerbated with the transition from a 2D chip system to a 3D stacked system (Sapatnekar 2009). The 3D integration can increase device density, bandwidth and speed. On the other hand, due to increased integration, the amount of heat per unit footprint increases, resulting in high on-chip temperature. This, in turn, degrades performance and reliability of the systems. Moreover, as NoC consists of different cores, each having its own power profile, area, frequency of operation etc., it results in non-uniform heating of the chip. Observation of the thermal contours of certain industrial chips shows that the temperature at the hotspots can really exceed $100 \, °C$ (Tsai et al. 2006). Due to the increased power density, heat removal is extremely important in the 3D IC (Banerjee et al. 2001) as the heat sink is often located far away from some of the layers. So, vertical through vias (*thermal vias*) as effective thermal conductors is an effective heat removal approach in the 3D IC (Goplen and Sapatnekar 2005; Cong and Zhang 2005; Li et al. 2006, 2008). To remove heat from stacked silicon layers and alleviate the hotspots at each layer, thermal vias create a heat flow path from silicon layers to the heat sink. However, thermal via consumes area which can be used otherwise for routing and/or transistors. Furthermore, it can increase the IC cost. So, its number ought to be constrained (Goplen and Sapatnekar 2005). The major challenge to remove heat from the 3D system is the distribution of the limited number of thermal vias in the die.

4   Testing of NoC-Based Multi-Core Systems

Testing an NoC-based system poses new challenges, compared to the bus-based SoCs and board-based designs. Here, test engineers are expected to reuse the NoC structure to transport the test data in parallel and thus reduce the testtime. Though this has the potential to reduce testtime; power consumption can go up significantly. Accordingly, many test scheduling strategies have been proposed in the literature for power-constrained testing of NoCs (Liu et al. 2005a). Moreover, satisfaction of power constraints does not necessarily imply thermal safety (Liu et al. 2005b; Rosinger et al. 2005) of the chip. The chip floorplan has a crucial role in determining the thermal behaviour of blocks within the IC. The test engineers generally do not have the liberty to alter the floorplan. Thus, thermal safety during testing of IC is done using the technique called scheduling. In this case, a test scheduling strategy must care of closely situated cores, in the NoC-based system, are not tested simultaneously, particularly if the power consumption of corresponding cores are also high.

It may be noted that the thermal safety during testing cannot be ignored for several reasons. First, power consumption during the test is quite high as compared to the normal mode of operation of the IC. The resultant increase in temperature will increase the leakage current, acting as an aid to further increase the power consumption, and thus leading to a thermal runaway. High local temperatures create local hotspots that may lead to burn-out and as a result yield loss. The variance in temperature across the IC changes the delay of different parts in the chip. This may cause some good ICs to fail delay test and/or some bad chips to pass, affecting the yield and quality of products. The major challenge to test an NoC-based system is to reduce the testtime while maintaining the thermal safety of the system.

In summary, the major challenges in design and test of NoC-based systems are the association of cores in the systems to the tasks/threads of an application. A proper association may alleviate the thermal problem in the 2D NoC-based system. Proper placement of limited vertical interconnects onto the router should be made for enhancing the performance of 3D NoC-based system, in addition to the proper distribution of limited thermal vias to remove excess heat. It is also desirable to test the NoC-based system quickly and at the same time ensuring thermal safety.

5 Issues in Multi-Core Systems Design with Integrated NoC and 3D Technologies

In NoC-based system design, while there exist quite a good number of research works on communication infrastructure design, communication methodology and evaluation framework, application mapping techniques have not been explored well. It is necessary to alleviate the heating problem in 2D NoC-based system and to improve the performance of 3D NoC-based system together with proper placement of a limited number of TSVs. Furthermore, the heat removal strategy in 3D NoC-based system using thermal via has not been studied well. Also, thermal-aware test process in NoC-based system needs to be explored. This book attempts to fill up the gaps by designing efficient mapping technique to alleviate the thermal problem in 2D NoC-based, mapping application tasks/threads together with TSV placement, including thermal via distribution to enhance the performance of 3D NoC-based system. Finally, the designed system has to be tested by the care of thermal safety to improve the yield.

6   Application Mapping and TSV Placement: A Combined Approach

An application consists of a set of tasks/threads, each of which is implemented by an IP-core. In this book, it has been assumed that application partitioning has been done at the core level and that the core is responsible to carry out a particular task/thread which has already been decided. Moreover, the programmability and other software aspects related to IP such as task clustering and scheduling are not considered. Thus, the application can be represented in the form of a *task graph* (Murali and Micheli 2004). The *task graph* of an application is a directed graph, $G(C, E)$, comprising of a set C of vertices (or tasks/threads) together with a set E of directed edges. The edge $e_{i,j} \epsilon E$ represents the communication between tasks/threads c_i and c_j. The bandwidth requirement between tasks/threads c_i and c_j is denoted by $comm_{i,j}$, the weight of edge $e_{i,j}$.

The NoC-based system topology can also be represented in the form of a *topology graph* (Murali and Micheli 2004). It is a directed graph $T(R, F)$ comprising of set R of vertices (or cores in the topology) together with a set F of directed edges. The edge $f_{i,j} \epsilon F$ represents the actual link between the routers/cores r_i and r_j. The weight of the edge $f_{i,j}$, represented as $bw_{i,j}$, denotes the bandwidth available across the edge $f_{i,j}$.

A mapping of the task graph $G(C, E)$ onto the topology graph $T(R, F)$ is represented by the function $\mathbb{M} : C \rightarrow R$, such that $\forall c_i \in C, \exists r_k \in R$ and $\mathbb{M}(c_i) = r_k$. The total communication cost of an application under this mapping decides the quality of mapping. A single commodity, d^k, $k = 1, 2, \ldots, |E|$ represents the communication between each pair of task. The quantity d^k corresponds to the communication between tasks c_i and c_j, that is, $comm_{i,j}$, the bandwidth requirement.

Let the quantity $x_{i,j}^k$ indicate the commodity d^k following the link (r_i, r_j). It is given by

$$x_{i,j}^k = \begin{cases} value(d^k), & \text{if link}(u_i, u_j) \epsilon \mathbb{P}(source(d^k), sink(d^k)) \\ 0, & \text{otherwise} \end{cases} \tag{1.1}$$

where $\mathbb{P}(m, n)$ indicates the routing path between the nodes m and n in the topology. To ensure the bandwidth limitation of individual links, the following constraint must be satisfied:

$$\sum_{k=1}^{|E|} x_{i,j}^k \leq bw_{i,j}, \quad \forall i, j \epsilon \{1, 2, \ldots, |R|\} \tag{1.2}$$

The communication cost T of a mapping solution is given by

$$T = \sum_{k=1}^{|E|} value(d^k) \times hopcount(source(d^k), \ sink(d^k)) \tag{1.3}$$

where, $hopcount(a, b)$ is the total number of hops between nodes (in the topology) a and b. For a deterministic shortest path routing, *hop count* corresponds to the minimum number of hops between the constituent nodes. Since *communication cost* is very much dependent on the mapping solution, the overall mapping problem is to optimize the *communication cost*, satisfying the bandwidth constraints of individual links. That is, the overall packet latency and network energy consumption are to be minimized, and the thermal safety of the system has to be ensured. The mapping of tasks to cores plays a crucial role in overall system performance. The approach corresponds to a design-time decision of attachment of tasks to cores.

From a theoretical viewpoint, the application mapping problem attempts to freeze one graph (the core graph) into another (the topology graph). This is intractable (Garcy and Johson 1979) and is a specific case of quadratic assignment problem (QAP) (Congying et al. 2011). Furthermore, proper placement of TSVs into die is also intractable, being an instance of the uncapacitated facility location problem (Guner and Sevkli 2008).

7   Swarm Based Optimizer

In this works, Particle Swarm Optimization (PSO) has been used as the tool to optimize system performance. There are several issues that justify the choice of PSO. It provides several advantages over other optimization strategies, such as Simulated Annealing (SA), Genetic Algorithm (GA) and Ant Colony Optimization (ACO). PSO is a population-based stochastic search technique. Multiple solutions co-exist at any stage of the process, whereas, SA progresses with only one solution at any stage of optimization. More precisely, PSO searches the solution space in a multidirectional (better) fashion. Thus, it is expected to produce solutions closer to the optimal one than SA. Such a solution is also produced quickly, compared to GA and ACO with limited usage of population and resources. The current best solution of a generation broadcasts its experience to all other fellow particles directly to provide better and quicker convergence towards the optimal solution. More precisely, it combines the local search method with the global search method (guided search) to balance exploration and exploitation. Moreover, particles (solutions) do not die in PSO by any selection criteria. The local best of a particle remains attached with the particle and gets updated whenever a better solution configuration is identified by the particle. In GA, population moves together (unguided search) and some solutions are filtered out by the natural selection criteria. In addition to this, the topology in GA varies as generation proceeds which can disturb the convergence. Similarly, in ACO, the entire memory of the colony is preserved, instead of only the previous generation. As random paths are selected with the help of colony memory for an ant, the solution takes time to converge.

8 Scope and Motivation of the Works

In designing 3D NoC-based system with a limited number of TSVs and maintaining a proper distance between the neighbouring TSVs, locations of TSVs play a major role. Such constraints make NoC routers asymmetric. Some routers have vertical connections (TSV) while others do not. Such 3D NoC-based systems are partially connected in the vertical direction (whereas, in fully connected one, every router has a vertical connection). In this topology, the average distance between cores is significantly more than the fully connected 3D NoC. Furthermore, it provides less bisection width compared to a fully connected one. In such a scenario, the performance of the system is guided by the proper association between tasks of the application to cores (known as application mapping) and positions of the TSVs. Various avenues have been explored to solve this application mapping problem together with the placement of TSVs. It includes some heuristic approaches and an approach based on PSO. System chip design must also consider the thermal effect. In this direction, this work proposes a PSO-based meta-heuristic to meet the thermal (peak temperature) constraints in two situations—with or without thermal vias. However, in 2D NoCs, the thermal problem can be mitigated by using a thermal-aware (application) mapping technique only. The PSO-based strategy has been augmented several ways to alleviate the thermal problem with some sacrifice in communication cost.

To test such NoC-based system while ensuring thermal safety, the thesis proposes two thermal-aware test scheduling strategies—non-preemptive and preemptive. The schedule generation procedures use PSO with several augmentations to find good solutions. Thermal safety has been assured with some sacrifice in testtime.

9 Summary of This Book

Significant research is being performed to design and test of NoC-based multi-core systems. Existing techniques use different approaches to solve this problem. The basic objective of this book is little different, where six interrelated research issues have been addressed to strive for a generic approach to design and test of NoC-based multi-core systems:

(a) Application mapping together with TSV placement algorithm using Kernighan–Lin (KL) partitioning technique has been proposed, for 3D NoC-based systems design, to reduce the communication cost.

(b) Constructive mapping together with TSV placement algorithm, based on prediction, has been proposed for 3D NoC-based systems design, to reduce the communication cost.

(c) A Discrete Particle Swarm Optimization (DPSO) based approach has been presented for application mapping together with TSV placement to minimize the overall communication cost. A high-quality random number generator has been used to generate random initial solutions. Next, the basic PSO has been extended to have multiple swarms. For any stage of PSO, the initial population generation is not fully random. A good number of particles have been created using a fast, deterministic search. Furthermore, each particle gets mutated after certain generations. This has enabled our PSO to explore the promising regions of search space much better. Next, to check the optimality of the proposed solution, an ILP has been formulated for application mapping together with TSV placement for 3D NoC with two layers. The communication cost metric values of the mapping together with TSV placement solutions of our approaches have been compared with existing approaches. These show good improvement in both solution quality and execution time of the mapping techniques. Comparison of dynamic performance (in terms of average network latency) and energy consumption has also been carried out.

(d) A DPSO based approach has been presented for thermal-aware mapping to ensure thermal safety with a nominal sacrifice in communication cost. Augmentations similar to the DPSO proposed in (c) have been incorporated to get good results. An ILP has also been formulated for thermal-aware mapping problem to see the optimality of the proposed solution for 2D NoC-based systems.

(e) A PSO-based approach has been presented to solve the thermal problem in the 3D NoC. It aims to meet the thermal constraint with or without deploying the thermal vias. To meet the thermal constraint without deploying the thermal vias, the approach tries to generate the solution by giving some tolerance in communication cost. While the other approach tries to deploy the thermal vias to meet the thermal constraint of the solution by providing some extra area in the die (i.e. tolerance on die area).

(f) To test such an NoC-based system, PSO-based non-preemptive and preemptive test scheduling approaches have been proposed to minimize the testtime while maintaining thermal safety. Next, to see the optimality of the proposed preemptive solution, an ILP has been formulated.

10 Conclusion

This chapter focuses on the challenges of traditional bus-based SoC design with increasing number of cores in DSM technology and also explains the necessity of NoC as new communication backbone. NoC provides a scalable SoC platform and also narrows down the design-productivity gap by efficient reuse of large number of IP-cores. Furthermore, the emerging 3D and NoC technologies together has been proposed to alleviate interconnect scaling demand. The chapter includes the research directions of this new paradigm. A brief discussion about TSV placement together with application mapping and thermal safety of such system including test process of such a system. The objective, motivation, and contribution of this book have been clearly stated. To start with, Chap. 2 presents the related works on application mapping together with TSV placement, thermal-aware design and test strategies available in the literature and describes different aspects of it.

References

Banerjee, K., Souri, S. J., Kapur, P., & Saraswat, K. C. (2001). 3-D ICs: A novel chip design for improving deep-submicrometer interconnect performance and systems-on-chip integration. *Proceedings of the IEEE, 89*(5), 602–633.

Benini, L., & De Micheli, G. (2002). Networks-on-chips: A new SoC paradigm. *Computer, 35*(1), 70–78.

Bjerregaard, T., & Mahadevan, S. (2006). A survey of research and practices of network-on-chip. *ACM Computing Surveys, 38*(1), 1–51.

Cong, J., & Zhang, Y. (2005). Thermal via planning for 3-D ICs. In *Proceedings of the International Conference on Computer-Aided Design (ICCAD)* (pp. 745–752). Piscataway, NJ: IEEE.

Congying, L., Huanping, Z., & Xinfeng, Y. (2011). Particle swarm optimization algorithm for quadratic assignment problem. In *Proceedings of IEEE International Conference on Computer Science Networking Technolgy* (pp. 728–1731). Piscataway, NJ: IEEE.

Dally, W. J., & Towles, B. (2001). Route packets, not wires: On-chip interconnection networks. In *Proceedings of Design Automation Conference (DAC)* (pp. 683–689).

Davis, W. R., Wilson, J., Mick, S., Xu, J., Hua, H., Mineo, C., et al. (2005). Demystifying 3D ICs: The pros and cons of going vertical. *IEEE Design and Test of Computers, 22*, 498–510.

FlipChip. (2005). *Flip chip ball grid array package reference guide.* www.ti.com/lit/ug/spru811a/spru811a.pdf

Garcy, M. R., & Johnson, D. S. (1979). *Computers and intractability: A guide to the theory of NP-completeness.* San Francisco, CA: W. H. Freeman Publisher.

Goplen, B., & Sapatnekar, S. (2005). Thermal via placement in 3D ICs. In *Proceedings of International Symposium on Physical Design (ISPD)* (pp. 167–174). New York, NY: ACM.

Grecu, C., Pande, P. P., Ivanov, A., & Saleh, R. (2004). Structured interconnect architecture: A solution for the non-scalability of bus-based SoCs. In *Proceedings of Great Lakes Symposium on VLSI (GLSVLSI)* (pp. 192–195). New York, NY: ACM.

Guner, A. R., & Sevkli, M. (2008). A discrete particle swarm optimization algorithm for uncapacited facility location problem. *Journal of Artificial Evolution and Application, 2008*, 1–9

Ho, R., Mai, K. W., & Horowitz, M. A. (2001). The future of wires. *Proceedings of the IEEE, 89*(4), 490–504.

Li, X., Ma, Y., Hong, X., Dong, S., & Cong, J. (2008). LP based white space redistribution for thermal via planning and performance optimization in 3D ICs. In *Proceedings of Asia and South Pacific Design Automation Conference (ASP-DAC)* (pp. 209–212). Piscataway, NJ: IEEE.

Li, Z., Hong, X., Zhou, Q., Zeng, S., Bian, J., Yang, H., et al. (2006). Integrating dynamic thermal via planning with 3D floorplanning algorithm. In *Proceedings of International Symposium on Physical Design (ISPD)* (pp. 178–185). New York, NY: ACM.

Liu, C., Shi, J., Cota, E., & Iyengar, V. (2005a). Power-aware test scheduling in network-on-chip using variable-rate on-chip clocking. In *Proceedings of VLSI Test Symposium* (pp 349–354).

Liu, C., Veeraraghavan, K., & Iyengar, V. (2005b). Thermal-aware test scheduling and hot spot temperature minimization for core based systems. In *Proceedings of International Symposium on Defect and Fault Tolerance in VLSI Systems (DFT)* (pp. 552–560).

Murali, S., & Micheli, G. D. (2004). Bandwidth constrained mapping of cores onto NoC architectures. In *Proceedings of Design, Automation and Test in Europe (DATE)* (pp. 896–901).

Pasricha, S. (2012). A Framework for TSV serialization-aware synthesis of application specific 3D networks-on-chip. In *Proceedings of International Conference on VLSI Design (VLSID)* (pp. 268–273).

Quaye, C. A. (2005). Thermal-aware mapping and placement for 3-D NoC design. In *Proceedings of IEEE International Conference on SoC* (pp. 25–28).

Rosinger, P., Al-Hashimi, B., & Chakrabarty, K. (2005). Rapid generation of thermal-safe test schedules. In *Proceedings of Design, Automation and Test in Europe (DATE)* (pp. 840–845).

Sapatnekar, S. S. (2009). Addressing thermal and power delivery bottlenecks in 3D circuits. In *Proceedings of Asia and South Pacific Design Automation Conference (ASP-DAC)* (pp. 423–428).

Semiconductor Industry Association. (2007). In *The international technology roadmap for semiconductors (ITRS)*.

Semiconductor Industry Association. (2009). In *The international technology roadmap for semiconductors (ITRS)*.

Shang, L., Peh, L., Kumar, A., & Jha, N. K. (2006). Temperature-aware on-chip networks. *IEEE Micro, 26*(1), 130–139.

Tsai, J. L., Chen, C. P., Chen, G., Goplen, B., Qian, H., Zhan, Y., et al. (2006). Temperature aware placement for SOCs. *Proceedings of the IEEE, 94*(8), 1502–1518.

Xu, T. C., Liljeberg, P., & Tenhunen, H. (2011). Optimal number and placement of through silicon vias in 3D network-on-chip. In *Proceedings of International Symposium on Design and Diagnostics of Electronic Circuits & Systems (DDECS)* (pp 105–110).

Chapter 2
Alternative Approaches

Network-on-Chip (NoC) has evolved as a standard to design multiprocessor based on-chip system (Goossens et al. 2005; STMicroelectronics 2005; Vangal et al. 2007; Jerger and Peh 2009). The holistic research problems in NoC domain can broadly be organized into five main categories (Marculescu et al. 2009). The first one is the choice of *communication infrastructure*, such as network topology, router/switch architecture, link design, clocking, buffer optimization, floorplanning and layout. The second dimension of NoC research deals with the *communication paradigm* including routing policies, switching strategies, congestion control, power and thermal management, fault-tolerance, reliability, etc. The third avenue includes *designing an evaluation framework* for NoC communication architecture to have a good understanding of achievable latency, energy/power consumption and bandwidth of the network. The next one, in overall system design, is to associate the tasks of an application with the cores, known as *application mapping* process. This has got a very significant role to play defining the overall system performance, including communication time, required link bandwidth and thermal safety. The final direction deals with *NoC design validation and synthesis*. That is, the chosen NoC design has to be prototyped, validated and tested. Several works have been carried out in each direction. This chapter looks into the different development in mapping techniques, including 3D NoC, thermal safety and test strategies in NoC-based systems.

Application mapping is one of the most important dimensions in NoC-based systems research. It maps the tasks/threads of the application to the cores of the NoC topology, affecting overall performance and thermal safety of the system.

A large number of research works on application mapping have been reported in recent years. This chapter studies these works and classifies them. In some NoCs, the cores are already fixed with the switches/routers. In such NoCs, task mapping technique attempts to decide which particular core to be used. However, in general NoCs, no such attachment pre-exists. Association between tasks/threads and cores is done in the application mapping phase. The IP-cores in a NoC topology can be of two types—*homogeneous* and *heterogeneous*. Cores with homogeneous nature

© Springer Nature Switzerland AG 2020
K. Manna, J. Mathew, *Design and Test Strategies for 2D/3D Integration for NoC-based Multicore Architectures*, https://doi.org/10.1007/978-3-030-31310-4_2

can perform a similar set of jobs while heterogeneous cores can perform different sets of jobs. This decision is taken in the core selection phase. In this book, it has been assumed that the core is responsible to carry out a particular task which has already been decided. Thus, the application can be represented in the form of a *task graph* (Murali and Micheli 2004a). In the mapping phase, task graph (including communication requirement and power consumption) comes as an input, and an association is established between such tasks and cores in the topology graph. More precisely, application mapping freezes task graph to topology graph which is an NP-hard problem (Murali and Micheli 2004a).

1 Application Mapping Techniques

The mapping process can be classified into two categories: *dynamic* and *static*. This classification is based on the time at which tasks are assigned to cores. In NoC, design-time (static) mapping is performed, as excessive communication demand, in dynamic mapping significantly affects system performance, increasing the overall delay of the system (Pop and Kumar 2004).

1.1 Dynamic Application Mapping Techniques

Dynamic or online mapping techniques take place at runtime. The ready tasks are mapped to the cores by observing the instantaneous activities of the cores. Thus, the association of tasks to cores can change, while executing the application.

For dynamic task mapping, a heuristic has been proposed in Carvalho et al. (2007) and de Souza Carvalho et al. (2010). Here, an initial task mapping is carried out prior to the dynamic task mapping. The dynamic mapping can be performed in several ways: *First Free* (FF), *Nearest Neighbour* (NN), *Minimum Maximum Channel Load* (MMC), *Minimum Average Channel Load* (MAC), *Path Load* (PL), etc. The FF strategy selects the first free IP-core in the network which can execute the required task. The free IP-core is searched in column-wise manner in the network. In NN mapping strategy, the required task is placed at the free neighbouring IP-core of the core making the request. Congestion has been taken care of in MMC by distributing the load to the links. The MAC approach is similar to MMC, which distributes the communication requirement onto the NoC to reduce the average load in the links. In such approach, every link in the NoC is observed before distributing the load which in turn increases the mapping time. Such problem is overcome in PL heuristic. This technique considers the links that are used by the task being mapped. Thus, it produces better solution compared to many other approaches. In Chou and Marculescu (2008b) the authors have presented an online strategy for task assignment to homogeneous NoC platform. Such an assignment has been performed by taking care of the user behaviour.

It allows the systems to respond better to adapt to the user needs dynamically and to real-time changes. A runtime agent-based distributed application mapping strategy for NoC-based heterogeneous MPSoCs has been described in Faruque et al. (2008). This strategy reduces the monitoring traffic and computational effort for mapping process, compared to the centralized approaches. An agent process can be executed on any node in the NoC to perform resource management and store state information of the resource. The agents communicate with each other to find the processing elements suitable for mapping a task. The process is performed using the cooperation of two types of agents: *Global Agents* (GA) and *Cluster Agent* (CA). The CAs have the information about their clusters. When a new task arrives at a cluster, the GAs negotiate with others having global information about all clusters. The MPSoC architecture with both software and hardware processing elements has been designed in Singh et al. (2009). In such an architecture, one core acts as a Manager processor (among the available processors) to support task mapping, task binding, task migration, reconfiguration control and resource control. The manager keeps track of the resource occupancy, mapping decisions are taken accordingly. The proposed mapping technique places the communicating tasks of an application close to each other to improve the performance. Furthermore, such a technique can recommend the adjacent tasks on the same processing element with the availability of resource to improve the performance, including task execution time. It works in two phases. In the first phase, initial task mapping is done either on the first free location found in the network that can support the task or the NoC grouped into cluster and the initial task is mapped to the centre of the cluster. In the second phase, the incoming tasks are placed for better performance gain by an efficient runtime mapping algorithm. A dynamic task mapping heuristic has been presented in Mandelli et al. (2011a,b) to reduce energy consumption, named low energy consumption-based dependencies-neighbourhood (LEC-DN). In such a technique, cost function includes distance in terms of hops and the communication volume among the tasks. LEC-DN uses the Nearest Neighbour (NN) search in spiral fashion when target task has only one communication task that has already mapped. On the other hand, if there are more than one communicating tasks that are already mapped, it looks for a processing element within the bounding box defined by the position of such tasks, depending on the communication volume. A resource-aware, application-driven and decentralized dynamic mapping has been reported in Weichslgartner et al. (2011) where tasks are embedded incrementally with the already mapped ones.

2 Static Application Mapping Techniques

In static mapping, the decision for task assignment is taken prior to the execution of the application. Such a decision is fixed throughout the execution and is taken off-line. More precisely, all the tasks are mapped to the cores at design-time. Several strategies have been proposed to find a good and efficient solution. The off-line mapping can be broadly categorized as *Exact mapping* and *Search based mapping*, based on the strategies applied to reach a mapping solution.

2.1 Exact Application Mapping Techniques

The optimal mapping solution can be found by the mathematical formulation of the problem. A Mixed Integer Linear Programming (MLIP) formulation for mapping tasks onto NoC has been reported in Rhee et al. (2004). It considers the placement of tasks and cores. The authors have evaluated the proposed approach using some real and random benchmarks. The results show energy reduction as compared to other approaches. In Murali and Micheli (2004b) the authors have reported an integrated approach for mapping tasks onto heterogeneous processor/memory-based NoC and physical planning. The size and position of cores and network components are computed. To place tasks onto the cores of NoC, initially, authors have used a greedy mapping of tasks onto specified topology. Thereafter, in the improvement phase, the relative tasks positions are fixed by *Tabu Search* (TS). In the final phase, it formulates an MILP-based physical planning to improve the power and area of the final design of an application, including *Quality-of-Service* (QoS). An MILP-based unified approach has been reported in Ghosh et al. (2009) for application mapping which reduces the energy consumption. Such an approach has also considered the other problems like operating voltage assignment and routing. In Huang et al. (2011) the authors have extended the existing ILP (Ghosh et al. 2009) to find a trade-off between communication and computation energy. A contention-aware application mapping has been reported in Chou and Marculescu (2008a) using an ILP formulation for the problem. The experimental results show the improvement in packet latency, but, the loss of communication energy is high. In Tosun et al. (2009) the authors have described an ILP formulation for application mapping on to mesh-based NoC to reduce the energy consumption for different benchmarks. However, it does not consider bandwidth constraints. It is seen that the CPU time reported in this work for different benchmarks is quite high. The authors have reported an approach to reduce such high CPU time in Tosun (2011a). In this approach, tasks are clustered, as in Srinivasan et al. (2006). The entire mesh is divided into smaller sized meshes. Each small meshes are considered as topology and tasks are mapped onto the cluster using the ILP in Tosun et al. (2009). In the end, all sub-meshes are merged together for the solution. Experimental results show the improvement in CPU time reduction with the sacrifice in communication cost of mapping solution.

2.2 Search-Based Application Mapping Techniques

Based on the search procedure and results, two categories of search-based mapping techniques are available: *deterministic* search and *heuristic* search.

2.2.1 Deterministic Search-Based Application Mapping Techniques

Search techniques using *Branch-and-Bound* (BB) belong to this category. In this search technique, the mapping solution is searched in a topological fashion of the tree branches and bounding unallowable solutions. Such a technique can be applied to small problems, as search time grows exponentially with the size of the problem. An energy- and performance-aware mapping has been noted in Hu and Marculescu (2003, 2005) to satisfy the design constraints through bandwidth reservation. The authors have first formulated *Energy and performance Mapping* (EPAM or GMAP) including efficient *Performance-aware Branch-and-Bound* (PBB) algorithm to improve the solution quality. The experimental results show significant energy saving compared to Simulated Annealing (SA) based approach (Hu and Marculescu 2003, 2005). To reduce the hotspot routers in NoC, the traffic balanced IP mapping algorithm (TBMAP) for mesh-based NoC has been reported in Lin et al. (2008). In this technique, traffic of all the routers is balanced without sacrificing the network performance. The authors have modified the Branch-and-Bound technique proposed in Hu and Marculescu (2003, 2005) to map all tasks. Branch-and-Bound algorithms take a high amount of memory space and suffer from long CPU time demands.

2.2.2 Heuristic Search-Based Application Mapping Techniques

In this approach, an existing mapping solution is transformed to arrive at a better one. Typical examples include evolutionary techniques, such as Genetic Algorithm (GA), Ant Colony Optimization (ACO) and Particle Swarm Optimization (PSO).

(a) **GA-based heuristic search techniques**

A Genetic Algorithm (GA) based technique has been proposed in Lei and Kumar (2003) for mapping applications onto NoC-based systems. It works in two steps to reduce the execution time. At first, the tasks are mapped onto different IPs assuming the edge delays to be equal to the average edge delay. In the second step, the IPs are assigned to the tiles of NoC considering actual edge delay based on the traffic model to minimize the overall system delay. In Bhardwaj and Jena (2009) the authors have proposed a multi-objective Genetic Algorithm (MOGA) based approach for application mapping problem. This technique minimizes the energy consumption and required bandwidth using one-one as well as the many-many mapping between switches and tiles.

CGMAP (Darbari et al. 2009a,b), a GA-based mapping strategy uses the chaotic mapping operator instead of the random process of GA. The concept of the chaotic sequence has been combined with the Genetic Algorithm to achieve an optimal solution. The authors in Fard et al. (2009) have described another one-dimensional chaotic mapping technique onto NoC-based system to achieve a better solution. An evolutionary approach, GBMAP (Tavanpour et al. 2010), for application mapping has been reported. While mapping, it minimizes energy consumption and total bandwidth requirement of the NoC.

A two-phase GA-based mapping has been proposed in Fen and Ning (2010), named GAMR, to generate a deterministic deadlock free minimal routing path for each communication, to minimize the total communication energy and maximize link bandwidth utilization of the NoC architecture. In this technique, at first phase, different resource nodes are placed onto NoC. In the next phase, it produces several deadlock free minimal routing path for each communication without disturbing the placement of core. A Genetic Algorithm-based mapping technique has been proposed in Choudhary et al. (2010) to reduce the communication cost for customized the NoC architecture. The same authors (Choudhary et al. 2011) described a GA-based mapping technique for irregular NoC topology. While mapping tasks onto cores of the NoC, it takes care of a reduction in communication energy. A multi-objective Adaptive Immune Algorithm (MAIA) based on evolutionary technique has been reported in Sepulveda et al. (2009) to map the tasks onto the cores of NoC, minimizing the power consumption and overall network latency. It incorporates several features to improve local search while preventing premature convergence by preserving the diversity of the population. An improved version of MAIA has been proposed in Sepulveda et al. (2011) to map multi-applications onto the cores of NoC-based systems. It produces a set of alternatives by exploring the mapping space.

The main disadvantage of the genetic search strategy is that there is no guarantee of convergence. It often requires the GA to evolve a large number of generations to converge to a solution. In the end, the best solution is taken to be the solution to the mapping problem.

(b) **PSO and ACO-based heuristic search techniques**

In Fekr et al. (2010) the authors have described a Particle Swarm Optimization (PSO) based application mapping strategy for NoC-based system. However, no comparison has been done with the existing approaches. Thus, the benefit of such a mapping approach is not clear. Another, PSO-based mapping technique has been presented in Lei and Xiang (2010). It shows relative improvement over GA-based method. In Benyamina et al. (2010), a hybrid multi-objective strategy has been described for mapping tasks on to the cores of NoC-based systems. It uses Dijkstra's shortest path algorithm to find the shortest path among communicating cores to satisfy the bandwidth constraints. After that, a multi-objective Pareto-based PSO algorithm is used to improve the performance. To reduce the static and dynamic communication cost of mesh-based NoC, PSMAP, a PSO-based technique has been used in Sahu et al. (2011).

An Ant Colony Optimization (ACO) based algorithm has been reported in Wang et al. (2011b) for mapping tasks onto NoC to minimize the bandwidth requirement. The experimental results have been compared with random mapping technique.

2.2.3 Constructive Heuristics Techniques

In constructive heuristics-based technique, the final mapping solution is derived from the partial solutions. The partial solutions are generated sequentially. It can be classified into two categories: constructive heuristics with iterative improvement and constructive heuristics without iterative improvement. It produces solution faster as compared to transformative heuristics.

(a) **Constructive heuristics without the iterative improvement techniques**

In this approach, tasks of an application are mapped onto the cores of the NoC-based systems, one at a time. The tasks are selected in some predefined order. There will be no change in the position of the tasks after such placement. No optimization is applied upon the initial solution to arrive at a better solution.

SMAP (Saeidi et al. 2007) is a simulation-based technique for mapping the tasks on to the cores of mesh-based NoC-system. It performs application mapping together with task routing to improve the execution time and energy consumption. In this technique, the highest priority task is mapped at the centre of the mesh. The remaining tasks are mapped spirally to the boundaries of the mesh by mapping highly communicating tasks nearby.

An efficient binomial application mapping and optimization algorithm (BMAP) has been reported in Shen et al. (2007). It works in three phases—IP ranking, merging IP sets and refreshing IP set. The tasks are ordered based on the communication bandwidth between them. The bandwidth of a task is measured by summing up incoming and outgoing communication bandwidth. Thus, based on the task order, the highest communicating IP sets are merged, two at a time, in every iteration. Now, it recalculates the bandwidth requirements among the IP sets by taking each IP set as an individual IP, which refers to the IP set.

In Tavanpour et al. (2009) a chain-based mapping algorithm has been reported, named CAMAP. It maps the chain of connected tasks nearby to minimize the total communication cost and energy. A reliability-aware application mapping strategy, RMAP, has been proposed in Patooghy et al. (2010). In this technique, the application graph is divided into two sub-graphs for minimizing the traffic flow among the sets and maximizing the traffic flow within the sub-graph. Then one sub-graph is mapped onto the upper triangular portion of the topology and the other sub-graph onto the lower triangular portion. In Yang et al. (2010) tasks are mapped onto the topology in a tree-like fashion. All nodes and the links among the nodes of mesh-based NoC systems are abstracted as a tree. In this tree model, the root is taken to be the vertex having the highest communication volume. The children of the root node are the vertices which communicate with root and so on. In the mapping process, the root node is placed at the center of the mesh and traversal is done from the center towards the borders of the NoC.

To minimize energy consumption, a mapping technique has been reported in Tosun (2011b). It proposes a routing technique for the NoC-based systems. In this approach, a priority list for the tasks is maintained based on their total and average communication bandwidth requirements. Mapping is based on this priority list. If two tasks have the same priority, it randomly selects one. Here, the initial task is selected based on the priority list. Next, the task having the highest communication with the already mapped task is selected for mapping. It maps the initial task at several selected locations in the NoC. Thus, it generates a set of solutions for each position of the initial task. The remaining nodes are mapped on to the NoC according to the priority list. Finally, the best solution is chosen as the mapping candidate from the set of solutions.

(b) **Constructive heuristic with the iterative improvement techniques**
    In this case, the tasks of an application are mapped onto NoC-based systems one at a time based on the same tasks ordering criteria to generate an initial solution. Thereafter, it invokes an iterative improvement routine on the initial solution to find a better solution.

    In NAMP (Murali and Micheli 2004a), the application mapping has been done with minimum path routing for a mesh-based NoC, taking care of bandwidth constraint, minimizing the average communication delay. It works in three phases. In the initial phase, the task with maximum communication demands is mapped to a core with maximum neighbours. Next, the task with the highest communication, with the already mapped task, is selected for mapping, and so on. In the next phase, Dijkstra's shortest path algorithm is used on the quadrant graph for finding the minimum path, with the satisfaction of bandwidth constraints. In the last phase, iterative improvement has been performed on the initial solution by invoking the minimum path computation procedure, for each pairwise swapping of the mapped tasks. It also includes a traffic splitting strategy into various paths of the mapping problem. A tool, SUNMAP, has been reported in Murali and Micheli (2004b) to automatically generate best standard topology for an application. It also performs the mapping tasks. The mapping considers the communication delay, area and power dissipation subject to bandwidth and area constraints. In LMAP (Sahu et al. 2010), the authors have proposed a mapping strategy based on Kerninghan-Lin (KL) bi-partitioning technique (Kernighan and Lin 1970). The proposed mapping reduces the static and dynamic communication cost when tasks are mapped onto NoC topology. It works in three phases. In the partition phase, that is, phase one, the application graph is partitioned using KL technique to find out the closeness of tasks by analyzing their bandwidth requirement. The second phase, called initial mapping, maps the tasks onto NoC-based systems, based on the final partitioning results produce in KL. Next, an iterative improvement phase is applied by swapping and flipping of cores within a partition to arrive at a final mapping solution.

3 Application Mapping Together with TSV Placement for 3D NoC-Based Multi-Core Systems

All the mapping techniques discussed earlier are applicable to 2D NoC-based systems. These techniques can also be extended for 3D NoC-based systems. However, in 3D NoC-based systems, vertical connections (TSVs) are limited due to the high area overhead in the die and also the creation of congestion in routing. Thus, it is required to limit the usage of such vertical connections. With the limited usage of TSVs, their placement in NoC plays a crucial role to improve the system performance. Therefore, mapping can be used to find out suitable locations for such limited TSVs.

The impact on the performance of 3D NoC due to the different number of TSVs and their placement has been analyzed in Xu et al. (2010, 2011). Here, authors have considered three types of TSV configurations, full, quarter and one-eighth connection and compared their performances. They have shown a trade-off between performance and manufacturing cost. The TSV squeezing scheme to share TSVs among neighbouring routers in 3D NoC-based systems has studied in Liu et al. (2011a). In this approach, four neighbouring routers share a TSV in a time division multiplexed fashion. In Hwang et al. (2011), a new communication technique has been proposed between TSVs. Here, four cores are taken to be in a single virtual group and one TSV is dedicated for the group. A core transfers traffic in the vertical direction by using the TSV either in its group or its neighbouring group, based on the current load of the TSV in the group.

Serialization of TSVs can contribute to reducing their number in 3D-IC (Pasricha 2009). TSV-virtualization also reduces the number of TSVs used in 3D-ICs (Miller et al. 2012). Here, the authors have proposed TSV-virtualization scheme for multi-protocol-interconnect in 3D-ICs. In this approach, TSVs are clocked at a much higher rate than conventional intra-layer links. To utilize the full bandwidth of TSV-based vertical interconnect, it uses multiple TSVs in a multiplexed and shared manner. Moreover, each layer can contain different types of interconnection architectures: buses, crossbars or NoCs.

An application-specific 3D NoC-based systems synthesis procedure has been presented in Yan and Lin (2008), based on a greedy rip-up-and-reroute technique. In this technique, smaller flows are routed first, followed by the larger flows. To reduce power consumption, a router merging scheme has been used to further optimize the topology. In Seiculescu et al. (2009), Murali et al. (2009), and Seiculescu et al. (2010) the authors have proposed a tool for NoC topology synthesis in 3D environment. It determines the custom topology for an application and paths for communication flows. Network components are assigned on to the 3D layer and placement of them is decided, in an individual layer. Here, TSVs are iteratively added during the synthesis process. To design custom 3D NoC-based systems for an application, a GA based synthesis procedure has been presented in Jiang and Watanabe (2010). The proposed procedure reduces the topology cost and optimizes the floorplan to reduce power consumption under software and hardware constraints.

A floorplan-aware 3D NoC-based systems synthesis technique has been described in Zhou et al. (2012). Here, the authors have used a Simulated Allocation (SAL), a stochastic method (multi-commodity flow technique), for traffic flow routing and also proposed a power and delay (queuing) model for network components. Also, they have presented a study of various factors having an impact on the network performance in 3D NoC-based systems, such as the number of TSVs and 3D tiers. An application-specific 3D NoC-based system design using ILP has been proposed in Xu et al. (2009). Here, the authors have considered low-radix routers and many long links. A framework, called MORPHEUS, for TSV serialization-aware synthesis of application-specific 3D NoC has been proposed in Pasricha (2012). It incorporates 3D topology, route generation and thermal-aware core layout. It also places the network interfaces (NIs), routers and serialized TSVs in the die. A high-level 3D NoC-based systems synthesis mechanism has been reported in Ying et al. (2012b) to improve system performance and to reduce the link load by distributing the communication evenly in the system. Here, a SA based algorithm has been used to optimize the overall system. In Zhong et al. (2011), the authors have presented a four-stage application-specific synthesis approach for 3D NoC. This approach attempts to generate power-performance efficient topology for an application. Task-to-task communication is analyzed from the task graph and it tries to place the most communicating pair in the same cluster. A TSV assignment procedure has also been incorporated to reduce the link power consumption. A GA-based optimization technique has been proposed in Ying et al. (2012a) to design 3D NoC-based systems with low vertical link density. It optimizes topology, routing algorithm, task mapping and tile placement. In Rahmani et al. (2012), the authors have proposed a bus-hybrid-symmetric-mesh based architecture for 3D NoC design. It is a combined version of the packet-switched network and bus-based communication. It enhances the system performance, thermal safety, fault-tolerance and power efficiency. In Kapadia and Pasricha (2012), the authors have proposed a framework for power delivery network (PDN) aware 3D-mesh-NoC-based systems synthesis with multiple voltage islands. They have proposed an ILP as well as a heuristic to synthesize the PDN. A Branch-and-Bound method has been used to generate multiple mappings to optimize NoC power. A co-synthesis methodology for PDN-3D-mesh-NoC-based systems has been reported in Kapadia and Pasricha (2013). To optimize the cost of PDN network and NoC-based systems design, they have used a bi-objective SA approach. An application-specific 3D-mesh-NoC-based system has been designed with traffic-aware selection strategy in Azampanah et al. (2013). The selection strategy can significantly balance the traffic load to reach a better performance. A latency-aware mapping scheme has been presented for regular 3D-mesh-NoC-based systems in Wang et al. (2011a). It uses a rank-based multi-objective GA to solve the mapping problem. Here, packet latency has been calculated in both congested and congestion free environments.

4 Thermal Management Techniques for NoC-Based Multi-Core Systems

The mapping techniques discussed so far do not consider the temperature effect during the mapping phase. Temperature affects the performance, power, and reliability of the system. A temperature-aware task mapping technique has been proposed in Xie and Hung (2006). It maps tasks using a heuristic and a floorplanning tool to reduce the peak temperature.

The power density in modern ICs has increased significantly in recent years and according to the ITRS, it is expected to increase further with the progress in VLSI technology, as there is relatively less progress in the reduction of operating voltage (Semiconductor Industry Association et al. 2003). A good number of works can be found in the literature related to the design of new packages to provide good heat removal capacity and improve the airflow on the circuit board. Circuits are packaged with die and kept against the spreader plate, generally made of a highly conductive material such as copper or aluminium, which is in turn put against copper or aluminium made, heat sink. The heat sink is further cooled by fan (Skadron et al. 2003).

Minimization of thermal hotspots can contribute to better thermal management at the system level. Qian and Tusi (Qian and Tsui 2011) have proposed a routing algorithm that works in runtime to reduce the thermal hotspots by distributing the traffic uniformly. An investigation of the CPU power level, hotspot location, hotspot size and local hotspot power density on its thermal performance has been carried out in Xu et al. (2004) and Xu (2006). A hotspot prevention strategy has been proposed in Link and Vijaykrishnan (2005) which balances the temperature by reconfiguring the functionalities at runtime across the IP-cores. The temperature can also be controlled by using a software-based approach, such as OS-level task scheduling which can be found in other kinds of literature for both single as well as multiprocessor systems. To get better thermal profile during normal operation, a thermal-aware runtime OS-level workload scheduling technique has been presented in Coskun et al. (2007). In addition, a heuristic is applied, when the temperature reaches a certain threshold, to distribute the workload in such a way that both spatial and temporal temperatures get reduced.

Several works can be found in the literature to control the temperature of IC during online/runtime and off-line, generally known as Dynamic Thermal Management (DTM) and Static Thermal Management (STM) technique, respectively. In MPSoC based system, DTM refers to software and hardware strategies which work dynamically, at runtime, to control the operating temperature of different IP-cores. However, in the STM technique, thermal management can be done beforehand, prior to the execution of the application, at the time of application mapping itself. Fetch toggling and computation migration are examples of DTM techniques (Skadron et al. 2003). ThermalHeard (Shang et al. 2004, 2006), a distributed runtime thermal management strategy, has been proposed to regulate the temperature profile of NoC-based system and ensure thermal safety during normal operation with a little impact on performance.

Static thermal management of MPSoC-based systems can be performed prior to the execution of the applications. In this approach, thermal safety can be ensured at the time of core placement, taking care of communication required between cores and also their temperature profile. The power consumed by the MPSoCs depends on several parameters, such as the physical location of cores, workload of individual cores and communication across the cores of the NoC. Moreover, all such parameters directly contribute to creating thermal hotspots (Quaye 2005). Hung et al. (2004) have described an IP-virtualization and placement algorithm based on GA for regular NoC architecture which attempts to achieve a good thermal behaviour of the system while reducing the on-chip communication using the judicious placement of IP-blocks. To minimize the average latency and achieve a thermal balance, a Pareto-based mapping technique using GA has been proposed in Zhou et al. (2006). A systematic design methodology has been proposed to reduce the peak temperature which is application independent (Anagnostopoulos et al. 2010). In Liu et al. (2011b), the authors have proposed a multi-objective ant-colony algorithm that maps tasks onto a 2D mesh based NoC architecture while taking care of energy consumption and temperature hotspots. A thermal-aware mapping technique, CoolMap (Moazzen et al. 2012), has been proposed to map an application onto mesh-based NoC systems that take care of thermal correlation among the IP-cores and the performance impact. Another thermal-aware mapping technique, TAPP, has been presented in Zhu et al. (2015) to reduce the peak temperature of the NoC with a little sacrifice in communication cost.

Several hybrid thermal management techniques are also available in the literature which uses both the strategies—STM and DTM, in a combined manner. In this approach, thermal balancing is done before the application is run as well as in the runtime to maintain the temperature of IP-cores. Maintenance of the temperature profile of MPSoC-based systems using such techniques can be found in Coskun et al. (2008). Here, an ILP-based technique has been applied to find out the static task schedule to minimize energy consumption and thermal hotspots. After that, an Operating System (OS)-level dynamic scheduling has applied to arrive at a better thermal condition, as in Coskun et al. (2007). To minimize the hotspot in MPSoC platform, a temperature-aware task mapping algorithm has been presented in Sarhan et al. (2010). To arrive at uniform thermal distribution, it uses the adaptive multi-threshold technique at runtime. In this approach, the proposed algorithm keeps monitoring the temperature of the cores and exchanges the tasks when the core temperature is relatively higher than the average circuit temperature. The cores may be turned off if they cross an absolute maximum temperature. For power reduction in NoC-based systems, different voltage-frequency selection techniques have also been proposed (Sarhan et al. 2010).

A thermal-aware mapping of 3D-mesh-NoC-based systems has presented in Hamedani et al. (2012). Here, an ILP-based approach and two heuristic-based static thermal-aware mapping algorithms have proposed to study the thermal constraints and their effects on temperature and performance. A mapping scheme has been proposed in Azampanah et al. (2013) to optimize the number of TSVs and peak temperature of 3D-symmetric-mesh-NoC-based systems. Here, the unused TSVs are kept as thermal TSV for heat dissipation.

5   Thermal-Aware Testing of NoC-Based Multi-Core Systems

Overheating problem during the test of NoC-based systems has been addressed in Liu and Iyengar (2006) and Liu et al. (2006). In these approaches, individual core temperature, during test, has been controlled by varying the test clock frequency assigned to each core, to achieve the thermal balance. The Works (Liu and Iyengar 2006; Liu et al. 2006) have proposed heuristics to achieve thermal optimization and reduce testtime by considering a thermal constraint. Another heuristic for thermal-safe test scheduling of SoC-based systems has been presented in He et al. (2007). In this approach, test sets have been divided into smaller test sub-sequences and cooling periods have been incorporated in between, such that, continuous application of test does not lead to a thermal violation. Further, the scheme interleaves the test sub-sequences from different test sets in such a way that a cooling period reserved for one core is utilized for the test transportation and application of another core. A thermal simulation driven test scheduling has been proposed in He et al. (2008). The main idea in this approach is to partition the test set. The novelty in this scheme is that instantaneous thermal simulation results are used to guide the partitioning process. Another partition-based SoC test scheduling algorithm has been proposed in Yao et al. (2011a). Here, power and thermal values have been taken as constraints. It partitions the test set and looks for partition having the earliest start time. To generate power and thermal profile, it uses the superposition principle instead of expensive thermal simulation. A thermal-aware *test-access mechanism* (TAM) design and test scheduling method for SoC has been presented in Yu et al. (2009). It is also based on test-set partitioning, interleaving and bandwidth matching. A thermal-cost model has been used to generate thermal profile by using cycle-accurate power profile for a schedule and maintain a thermal constraint. In He et al. (2009), thermal-aware test scheduling for core-based SoC, using an *Abort-on-First-Fail* (AOFF) test environment, has been described. In AOFF environment, the test period is terminated as soon as the first fault is detected. To avoid high temperature, test sets are partitioned into sub-sequences, including the cooling periods. Thermal simulation guides the test partitioning process. A temperature-aware SoC test scheduling, considering inter-chip process variation, has been proposed in Aghaee et al. (2010). In this work, two scheduling approaches have been proposed to maximize the test throughput, in the presence of inter-chip variation. The first approach is based on an extension of the traditional test scheduling algorithm. The second approach is based on the chip classification scheme and reading of a temperature sensor. A SA based thermal and power-aware test scheduling of cores in an NoC-based system using multiple clock rate has been presented in Salamy and Harmanani (2012). A dynamic thermal-aware test scheduling method using on-chip temperature sensors has been presented in Yao et al. (2011b). It also modifies the test architecture to support the dynamic thermal test scheduling technique.

6 Conclusion

This chapter surveys the application mapping strategies on to NoC-based systems, including 3D NoC-based systems with TSV placement and thermal-aware design. It also reports the works related to NoC-based testing, including 3D environment and thermal safety. It classifies the reported techniques into groups like dynamic and static mapping approaches. Static mapping techniques have further been categorized as exact methods, Branch-and-Bound, transformative and constructive approaches. Chapter 3 presents application mapping together with TSV placement techniques for 3D-mesh-based-NoC systems using Kernighan–Lin partitioning scheme along with an iterative improvement stage.

References

Aghaee, N., He, Z., Peng, Z., & Eles, P. (2010). Temperature-aware soc test scheduling considering inter-chip process variation. In *Proceedings of Asian Test Symposium (ATS)* (pp. 395–398).

Anagnostopoulos, I., Bartzas, A., & Soudris, D. (2010). Application specific temperature reduction systematic methodology for 2D and 3D network-on-chip. In *Proceedings of International Workshop on Power and Timing Modeling Optimization and Simulation (PATMOS)* (pp. 86–95).

Azampanah, S., Eskandari, A., Khademzadeh, A., & Karimi, F. (2013). Traffic-aware selection strategy for application-specific 3D NoC. *Advances in Computer Science: An International Journal, 2*(5), 107–117.

Benyamina, A. E. H., Boulet, P., Aroui, A., Eltar, S., & Dellal, K. (2010). Mapping real time applications on NoC architecture with hybrid multi-objective algorithm. In *Proceedings of International Conference on Metaheuristics and Nature Inspired Computing* (pp. 1–10).

Bhardwaj, K., Jena, R. K. (2009) Energy and bandwidth aware mapping of IPs onto regular NoC architectures using multi-objective genetic algorithms. In *Proceedings of International Symposium on System-on-Chip (SOC)* (pp. 27–31).

Carvalho, E., Calazans, N., & Moraes, F. (2007). Heuristics for dynamic task mapping in NoC-based heterogeneous MPSoCs. In *Proceedings of International Workshop on Rapid System Prototyping (RSP)* (pp. 34–40).

Chou, C. L., & Marculescu, R. (2008a). Contention-aware application mapping for network-on-chip communication architectures. In *Proceedings of International Conference on Computer Design (ICCD)* (pp. 164–169).

Chou, C. L., & Marculescu, R. (2008b). User-aware dynamic task allocation in network-on-chip. In *Proceedings of Design, Automation and Test in Europe (DATE)* (pp. 1232–1237).

Choudhary, N., Gaur, M. S., Laxmi, V., & Singh, V. (2010). Energy aware design methodologies for application specific NoC. In *Proceedings of NORCHIP* (pp. 1–4).

Choudhary, N., Gaur, M. S., Laxmi, V., & Singh, V. (2011). GA based congestion aware topology generation for application specific NoC. In *Proceedings of IEEE International Symposium on Electronics Design, Test, and Application* (pp. 93–98).

Coskun, A. K., Rosing, T. S., & Whisnant, K. (2007). Temperature aware task scheduling in MPSoCs. In *Proceedings of the Design, Automation and Test in Europe (DATE)* (pp. 1–6)

Coskun, A. K., Rosing, T. S., Whisnant, K. A., & Gross, K. C. (2008). Static and dynamic temperature-aware scheduling for multiprocessor SoCs. *IEEE Transactions on Very Large Scale Integration (VLSI) Systems, 16*(9), 1127–1140

Darbari, F. M., Khademzadeh, A., & Fard, G. G. (2009a). Evaluating the performance of a chaos genetic algorithm for solving the network on chip mapping problem. In *Proceedings of International Conference on Computational Science and Engineering* (pp. 366–373).

Darbari, F., Khademzade, A., & Gharooni-Fard, G. (2009b). CGMAP: A new approach to network-on-chip mapping problem. *IEICE Electronics Express, 6*(1), 27–34.

de Souza Carvalho, E. L., Calazans, N. L. V., & Moraes, F. G. (2010). Dynamic task mapping for MPSoCs. *IEEE Design Test of Computers, 27*(5), 26–35.

Fard, G. G., Khademzadeh, A., Darbari, F. M. (2009). Evaluating the performance of one-dimensional chaotic maps in network-on-chip mapping problem. *IEICE Electronics Express, 6*(12), 811–817.

Faruque, A., Abdullah, M., Krist, R., & Henkel, J. (2008). ADAM: Run-time agent based distributed application mapping for on-chip communication. In *Proceedings of Design Automation Conference (DAC)* (pp. 760–765).

Fekr, A. R., Khademzadeh, A., Janidarmian, M., & Bokharaei, V. S. (2010). Bandwidth/fault/contention aware application-specific NoC using PSO as a mapping generator. *World Congress on Engineering, 1*, 247–252.

Fen, G., & Ning, W. (2010). Genetic algorithm based mapping and routing approach for network-on-chip architectures. *Chinese Journal of Electronics, 19*(1), 91–96.

Ghosh, P., Sen, A., & Hall, A. (2009). Energy efficient application mapping to NoC processing elements operating at multiple voltage levels. In *Proceedings of International Symposium on Network-on-Chip (NoCS)* (pp. 80–85).

Goossens, K., Dielissen, J., & Radulescu, A. (2005). Athereal network on chip: Concepts, architectures, and implementations. *IEEE Design and Test of Computers, 22*(5), 414–421.

Hamedani, P. K., Hessabi, S., Sarbazi-Azad, H., & Jerger, N. E. (2012). Exploration of temperature constraints for thermal aware mapping of 3D networks-on-chip. In *Proceedings of Euromicro International Conference on Parallel, Distributed and Network-Based Processing (PDP)* (pp. 499–506).

He, Z., Peng, Z., & Eles, P. (2007). A heuristic for thermal-safe soc test scheduling. In *Proceedings of International Test Conference (ITC)* (pp. 1–10).

He, Z., Peng, Z., & Eles, P. (2008). Simulation-driven thermal-safe test time minimization for system-on-chip. In *Proceedings of Asian Test Symposium (ATS)* (pp. 283–288).

He, Z., Peng, Z., & Eles, P. (2009) Thermal-aware test scheduling for core-based soc in an abort-on-first-fail test environment. In *Proceedings of Euromicro Conference on Digital System Design* (pp. 239–246).

Hu, J., & Marculescu, R. (2003). Energy-aware mapping for tile-based NoC architectures under performance constraints. In *Proceedings of Asia and South Pacific Design Automation Conference (ASP-DAC)* (pp. 233–239)

Hu, J., & Marculescu, R. (2005). Energy- and performance-aware mapping for regular NoC architectures. *IEEE Transactions on Computer-Aided Design of Integrated Circuits and Systems, 24*(4), 551–562.

Huang, J., Buckl, C., Raabe, A., & Knoll, A. (2011). Energy-aware task allocation for network-on-chip based heterogeneous multiprocessor systems. In *Proceedings of Euromicro International Conference on Parallel, Distributed and Network Based Processing (PDP)* (pp. 447–454).

Hung, W., Quaye, C. A., Theocharides, T., Xie, Y., Vijaykrishnan, N., Irwin, M. J. (2004) Thermal-aware IP virtualization and placement for network-on-chip architecture. In *Proceedings of International Conference on Computer Design (ICCD)* (pp. 430–437).

Hwang, Y. J., Lee, J. H., & Han, T. H. (2011). 3D Network-on-Chip system communication using minimum number of TSVs. In *Proceedings of ICT Convergence (ICTC)* (pp. 517–522).

Jerger, E. N., & Peh, L. S. (2009). *On-chip networks*. San Rafael, CA: Morgan and Clay Pool Publisher.

Jiang, X., & Watanabe, T. (2010). An efficient 3D NoC synthesis by using genetic algorithms. In *Proceedings of IEEE Region 10 Conference TENCON* (pp. 1207–1212).

Kapadia, N., & Pasricha, S. (2012). A power delivery network aware framework for synthesis of 3D networks-on-chip with multiple voltage islands. In *Proceedings of International Conference on VLSI Design (VLSID)* (pp. 262–267).

Kapadia, N., & Pasricha, S. (2013). A co-synthesis methodology for power delivery and data interconnection networks in 3D ICs. In *Proceedings of International Symposium on Quality Electronic Design (ISQED)* (pp. 73–79).

Kernighan, B., & Lin, S. (1970). An efficient heuristic procedure for partitioning graphs. *Bell System Technical Journal, 49*(2), 291–307.

Lei, T., & Kumar, S. (2003). A two-step genetic algorithm for mapping task graphs to a network on chip architecture. In *Proceedings of Euromicro Symposium on Digital System Design (DSD)* (pp. 180–187).

Lei, W., & Xiang, L. (2010). Energy- and latency-aware NoC mapping based on discrete particle swarm optimization. In *Proceedings of International Conference on Communications and Mobile Computing* (pp. 263–268).

Lin, T. J., Lin, S. Y., & Wu, A. Y. (2008). Traffic-balanced IP mapping algorithm for 2D-mesh on-chip-networks. In *Proceedings of Workshop on Signal Processing Systems (SiPS)* (pp. 200–203).

Link, G. M., & Vijaykrishnan, N. (2005). Hotspot prevention through runtime reconfiguration in network-on-chip. In *Proceedings of the Design, Automation and Test in Europe (DATE)* (pp. 648–649).

Liu, C., & Iyengar, V. (2006). Test scheduling with thermal optimization for network-on-chip systems using variable-rate on-chip clocking. In *Proceedings of Design, Automation and Test in Europe (DATE)* (pp. 6–10).

Liu, C., Iyengar, V., & Pradhan, D. K. (2006). Thermal-aware testing of network-on-chip using multiple-frequency clocking. In *Proceedings of VLSI Test Symposium* (pp. 46–51).

Liu, C., Zhang, L., Han, Y., & Li, X. (2011a). Vertical interconnects squeezing in symmetric 3D mesh network-on-chip. In *Proceedings of Asia and South Pacific Design Automation Conference (ASP-DAC)* (pp. 357–362).

Liu, Y., Ruan, Y., Lai, Z., & Jing, W. (2011b). Energy and thermal aware mapping for mesh-based NoC architectures using multi-objective ant colony algorithm. In *Proceedings of International Conference on Computer Research and Development (ICCRD)* (pp. 407–411).

Mandelli, M., Amory, A., Ost, L., & Moraes, F. G. (2011a). Multi-task dynamic mapping onto NoC-based MPSoCs. In *Proceedings of Symposium on Integrated Circuits and System Design* (pp. 191–196).

Mandelli, M., Ost, L., Carara, E., Guindani, G., Gouvea, T., Medeiros, G., & Moraes, F. G. (2011b). Energy-aware dynamic task mapping for NoC-based MPSoCs. In *Proceedings of International Symposium on Circuits and Systems (ISCAS)* (pp. 1676–1679).

Marculescu, R., Ogras, U. Y., Peh, L. S., Jerger, N. E., & Hoskote, Y. (2009). Outstanding research problems in NoC design: System, microarchitecture, and circuit perspectives. *IEEE Transactions on Computer-Aided Design of Integrated Circuits and Systems, 28*(1), 3–21.

Miller, F., Wild, T., & Herkersdorf, A. (2012). TSV-virtualization for multi-protocol-interconnect in 3D-ICs. In *Proceedings of Euromicro Conference on Digital System Design (DSD)* (pp. 374–381).

Moazzen, M., Reza, A., & Reshadi, M. (2012). CoolMap: A thermal-aware mapping algorithm for application specific networks-on-chip. In *Proceedings of Euromicro Conference on Digital System Design (DSD)* (pp. 731–734). https://doi.org/10.1109/DSD.2012.35

Murali, S., & De Micheli, G. (2004a). Bandwidth constrained mapping of cores onto NoC architectures. In *Proceedings of Design, Automation and Test in Europe (DATE)* (pp. 896–901).

Murali, S., & De Micheli, G. (2004b). SUNMAP: A tool for automatic topology selection and generation for NoCs. In *Proceedings of Design Automation Conference* (DAC) (pp. 914–919)

Murali, S., Seiculescu, C., Benini, L., & De Micheli, G. (2009) Synthesis of networks on chips for 3D systems on chips. In *Proceedings of Asia and South Pacific Design Automation Conference (ASP-DAC)* (pp. 242–247). https://doi.org/10.1109/ASPDAC.2009.4796487

Pasricha, S. (2009). Exploring serial vertical interconnects for 3D ICs. In *Proceedings of Design Automation Conference (DAC)* (pp. 581–586).

Pasricha, S. (2012). A framework for TSV serialization-aware synthesis of application specific 3D networks-on-chip. In *Proceedings of International Conference on VLSI Design (VLSID)* (pp. 268–273).

Patooghy, A., Tabkhi, H., & Miremadi, S. G. (2010). RMAP: A reliability-aware application mapping for network-on-chips. In *Proceedings of International Conference on Dependability* (pp. 112–117).

Pop, R., & Kumar, S. (2004). A survey of techniques for mapping and scheduling applications to network on chip systems. School of Engineering, Jonkoping University, Research Report 4.4.

Qian, Z., & Tsui, C. (2011). A thermal-aware application specific routing algorithm for network-on-chip design. In *Proceedings of Asia and South Pacific Design Automation Conference (ASP-DAC)* (pp. 449–454).

Quaye, C. A. (2005). Thermal-aware mapping and placement for 3-D NoC design. In *Proceedings of IEEE International Conference on SoC* (pp. 25–28).

Rahmani, A. M., Vaddina, K. R., Latif, K., Liljeberg, P., Plosila, J., & Tenhunen, H. (2012). Design and management of high-performance, reliable and thermal-aware 3D networks-on-chip. *IET Circuits, Devices Systems, 6*(5), 308–321.

Rhee, C. E., Jeong, H. Y., & Ha, S. (2004). Many-to-many core-switch mapping in 2-D Mesh NoC architectures. In *Proceedings of International Conference on Computer Design: VLSI in Computers and Processors (ICCD)* (pp. 438–443).

Saeidi, S., Khademzadeh, A., & Mehran, A. (2007). SMAP: An intelligent mapping tool for network on chip. In *Proceedings of International Symposium on Signals, Circuits and Systems (ISSCS)* (pp. 1–4).

Sahu, P. K., Shah, N., Manna, K., & Chattopadhyay, S. (2010). A new application mapping algorithm for mesh based network-on-chip design. In *Proceedings of India Conference (INDICON)* (pp. 1–4).

Sahu, P. K., Venkatesh, P., Gollapalli, S., & Chattopadhyay, S. (2011). Application mapping onto mesh structured network-on-chip using particle swarm optimization. In *Proceedings of Computer Society Annual Symposium on VLSI (ISVLSI)* (pp. 335–336).

Salamy, H., & Harmanani, H. (2012). An effective solution to thermal-aware test scheduling on network-on-chip using multiple clock rates. In *Proceedings of Midwest Symposium on Circuits and Systems (MWSCAS)* (pp. 530–533).

Sarhan, H., Eddash, O. K., Raymond, M., Wassal, A., & Ismail, Y. (2010). Temperature-aware adaptive task-mapping targeting uniform thermal distribution in MPSoC platforms. In *Proceedings of International Conference on Energy-Aware Computing (ICEAC)* (pp. 1–3).

Seiculescu, C., Murali, S., Benini, L., & De Micheli, G. (2009). SunFloor 3D: A tool for networks on chip topology synthesis for 3D systems on chips. In *Proceedings of Design Automation Test in Europe (DATE)* (pp. 9–14).

Seiculescu, C., Murali, S., Benini, L., & De Micheli, G. (2010). SunFloor 3D: A tool for networks on chip topology synthesis for 3-d systems on chips. *IEEE Transactions on Computer-Aided Design of Integrated Circuits and Systems, 29*(12), 1987–2000.

Semiconductor Industry Association, et al. (2003). *The International Technology Roadmap for Semiconductors (ITRS)*.

Sepulveda, M. J., Strum, M., & Chau, W. J. (2009). A multi-objective adaptive immune algorithm for NoC mapping. In *Proceedings of International Conference on Very Large Scale Integration (VLSI-SOC)* (pp. 193–196).

Sepulveda, J., Strum, M., Chau, W. J., & Gogniat, G. (2011). A multi-objective approach for multi-application NoC mapping. In *Proceedings of Latin American Symposium on Circuits and Systems (LASCAS)* (pp. 1–4).

Shang, L., Peh, L. S., Kumar, A., & Jha, N. K. (2004). Thermal modeling, characterization and management of on-chip networks. In *Proceedings of International Symposium on Microarchitecture* (pp. 67–78).

Shang, L., Peh, L., Kumar, A., & Jha, N. K. (2006). Temperature-aware on-chip networks. *IEEE Micro, 26*(1), 130–139.

Shen, W. T., Chao, C. H., Lien, Y. K., & Wu, A. Y. (2007). A new binomial mapping and optimization algorithm for reduced-complexity mesh-based on-chip network. In *Proceedings of International Symposium on Networks-on-Chip (NoCS)* (pp. 317–322).

Singh, A. K., Jigang, W., Prakash, A., & Srikanthan, T. (2009). Mapping algorithms for NoC-based heterogeneous MPSoC platforms. In *Proceedings of Euromicro Conference on Digital System Design (DSD)* (pp. 133–140).

Skadron, K., Stan, M. R., Huang, W., Velusamy, S., Sankaranarayanan, K., & Tarjan, D. (2003). Temperature-aware microarchitecture. In *Proceedings of IEEE International Symposiun on Computer Architecture (ISCA)* (pp. 1–12).

Srinivasan, K., Chatha, K. S., & Konjevod, G. (2006). Linear-programming-based techniques for synthesis of network-on-chip architectures. *IEEE Transactions on Very Large Scale Integration (VLSI) Systems, 14*(4), 407–420.

STMicroelectronics. (2005). STNoC: Building a new system-on-chip paradigm. White paper.

Tavanpour, M., Khademzadeh, A., & Janidarmian, M. (2009). Chain-mapping for mesh based network-on-chip architecture. *IEICE Electronics Express, 6*(22), 1535–1541.

Tavanpour, M., Khademzadeh, A., Pourkiani, S., & Yaghobi, M. (2010). GBMAP: An evolutionary approach to mapping cores onto a mesh-based NoC architecture. *Journal of Communication and Computer, 7*(3), 1–7.

Tosun, S. (2011a). Clustered-based application mapping method for network-on-chip. *Journal of Advances in Engineering Software, 42*(10), 868–874.

Tosun, S. (2011b). New heuristic algorithm for energy aware application mapping and routing on mesh-based NoCs. *Journal of System Architecture, 57*, 69–78.

Tosun S, Ozturk O, Ozen M (2009). An ILP formulation for application mapping onto network-on-chips. In *Proceedings of International Conference on Application of Information and Communication Technologies (AICT)* (pp. 1–5).

Vangal, S., Howard, J., Ruhl, G., Dighe, S., Wilson, H., Tschanz, J., et al. (2007). An 80-Tile 1.28TFLOPS network-on-chip in 65 nm CMOS. In *Proceedings of International Solid-State Circuits Conference (ISSCC)* (pp. 98–589).

Wang, J., Li, L., Pan, H., He, S., & Zhang, R. (2011a). Latency-aware mapping for 3D NoC using rank-based multi-objective genetic algorithm. In *Proceedings of International Conference on ASIC (ASICON)* (pp. 413–416).

Wang, J., Li, Y., Chai, S., & Peng, Q. (2011b). Bandwidth-aware application mapping for NoC-based MPSoCs. *Journal of Computational Information Systems, 7*(1), 152–159.

Weichslgartner, A., Wildermann, S., & Teich, J. (2011). Dynamic decentralized mapping of tree-structured applications on NoC architectures. In *Proceedings of International Symposium on Network-on-Chip (NoCS)* (pp. 201–208).

Xie, Y., & Hung, W. L. (2006). Temperature-aware task allocation and scheduling for embedded multiprocessor system-on-chip (MPSoC) design. *Journal of VLSI Signal Processing, 45*, 177–189.

Xu, G. (2006). Thermal modeling of multi-core processors. In *Proceedings of Ninth Intersociety Conference on Thermal and Thermomechanical Phenomena in Electronics Systems (ITHERM)* (pp. 96–100).

Xu, G., Guenin, B., & Vogel, M. (2004). Extension of air cooling for high power processors. In *Proceedings of Intersociety Conference on Thermal and Thermomechanical Phenomena in Electronic Systems (ITHERM)* (pp. 186–193).

Xu, T. C., Liljeberg, P., Tenhunen, H. (2010) A study of through silicon via impact to 3D network-on-chip design. In *Proceedings of International Conference on Electronics and Information Engineering (ICEIE)* (pp. 333–337).

Xu, T. C., Liljeberg, P., & Tenhunen, H. (2011). Optimal number and placement of through silicon vias in 3D network-on-chip. In *Proceedings of International Symposium on Design and Diagnostics of Electronic Circuits & Systems (DDECS)* (pp. 105–110).

Xu, Y., Du, Y., Zhao, B., Zhou, X., Zhang, Y., & Yang, J. (2009). A low-radix and low-diameter 3D interconnection network design. In *Proceedings of Symposium on High Performance Computer Architecture (HPCA)* (pp. 30–42).

Yan, S., & Lin, B. (2008). Design of application-specific 3D networks-on-chip architectures. In *Proceedings of International Conference on Computer Design (ICCD)* (pp. 142–149).

Yang, B., Xu, T. C., Santti, T., & Plosila, J. (2010). Tree-model based mapping for energy-efficient and low-latency network-on-chip. In *Proceedings of International Symposium on Design and Diagnostics of Electronics Circuits and Systems (DDECS)* (pp. 189–192).

Yao, C., Saluja, K. K., & Ramanathan, P. (2011a). Power and thermal constrained test scheduling under deep submicron technologies. *IEEE Transactions on Computer-Aided Design of Integrated Circuits and Systems, 30*, 317–322.

Yao, C., Saluja, K. K., & Ramanathan, P. (2011b). Thermal-aware test scheduling using on-chip temperature sensors. In *Proceedings of VLSI Design (VLSID)* (pp. 376–381).

Ying, H., Heid, K., Hollstein, T., & Hofmann, K. (2012a). A genetic algorithm based optimization method for low vertical link density 3-dimensional networks-on-chip many core systems. In *Proceedings of NORCHIP* (pp. 1–4).

Ying, H., Hollstein, T., & Hofmann, K. (2012b). Communication-centric high level synthesis metrics for low vertical channel density 3-dimensional networks-on-chip. In *Proceedings of International Workshop on Reconfigurable Communication-Centric Systems-on-Chip (ReCoSoC)* (pp. 1–6).

Yu, T. E., Yoneda, T., Chakrabarty, K., & Fujiwara, H. (2009). Test infrastructure design for core-based system-on-chip under cycle-accurate thermal constraints. In *Proceedings of Design Automation Conference (DAC)* (pp. 793–798).

Zhong, W., Chen, S., Ma, F., Yoshimura, T., & Goto, S. (2011). Floorplanning driven network-on-chip synthesis for 3-D SoCs. In *Proceedings of International Symposium on Circuits and Systems (ISCAS)* (pp. 1203–1206).

Zhou, P., Yuh, P. H., & Sapatnekar, S. S. (2012). Optimized 3D network-on-chip design using simulated allocation. *ACM Transactions on Design Automation of Electronic Systems, 17*(2), 12:1–12:19.

Zhou, W., Zhang, Y., & Mao, Z. (2006). Pareto based multi-objective mapping IP cores onto NoC architecture. In *Proceedings of Asia Pacific Conference on Circuits and System (APCCAS)* (pp. 331–334).

Zhu, D., Chen, L., Pinkston, T. M., & Pedram, M. (2015). TAPP: Temperature-aware application mapping for NoC-based many-core processors. In *Proceedings of Design, Automation Test in Europe (DATE)* (pp. 1241–1244).

Chapter 3
Iterative Application Mapping with TSV Placement Strategy for Design a 3D NoC-Based Multi-Core Systems

Iterative mapping algorithms start with an initial mapping solution which is improved further by introducing changes into it. The problem can be viewed as the partitioning problem in VLSI physical design process. In a partitioning process, highly connected modules are put into the same partition to reduce the wiring overhead. Similarly, a mapping process has the objective to keep the highly communicating tasks close to each other.

Borrowing the concepts from physical design, this chapter presents application mapping strategies built around the Kernighan–Lin (KL) partitioning technique (Kernighan and Lin 1970). The KL-partitioning ensures that the tasks that communicate with each other frequently are within the same partition. An initial mapping onto NoC topology has been done using the final result of the KL-partitioning algorithm. An iterative improvement technique has next been applied to this initial mapping to improve the communication cost further. The technique consists of swapping and flipping the mapped task positions, including vertical direction, to arrive at a better solution (Fig. 3.1).

The proposed solution strategy works in two phases. In the first phase, it is assumed that all routers in the NoC have vertical connections (i.e., 3D router with a TSV). The mapping algorithm works on this fully connected 3D mesh-based NoC systems. Thereafter, it preserves those TSVs which carry the maximum traffic, based on the given constraints; and remove the remaining TSVs. Then, it maps the application on this vertically-partially-connected 3D mesh-based NoC systems. The salient features of the proposed approach are as follows:

1. The KL-based application mapping algorithm has been extended for 3D NoC-based systems together with TSV placement. It also incorporates several improvement operations on the initial mapping solution to reduce the communication cost.
2. The communication cost metric values of the mapping solutions of the approach have been compared with other existing strategies.

© Springer Nature Switzerland AG 2020
K. Manna, J. Mathew, *Design and Test Strategies for 2D/3D Integration for NoC-based Multicore Architectures*, https://doi.org/10.1007/978-3-030-31310-4_3

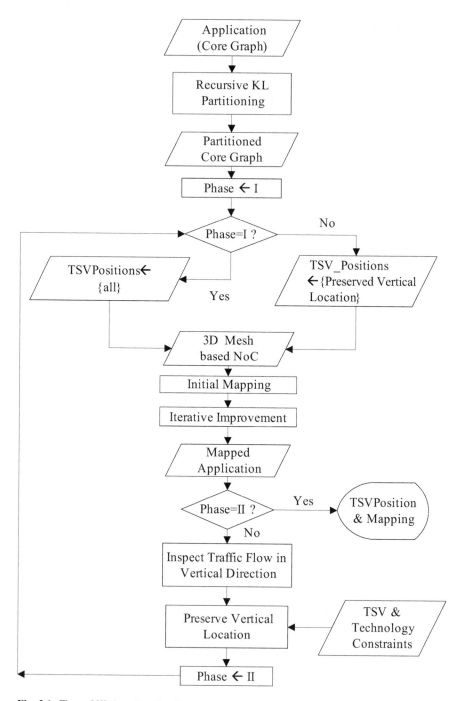

Fig. 3.1 Flow of KL-based application mapping and TSV placement for 3D NoC-based systems

3. Comparison of dynamic performance (in terms of average network latency) and energy consumption have also been carried out.

1 3D NoC-Based Systems and Routing Algorithm

A simple way to extend 2D mesh-based NoC systems to 3D is to extend each and every router in the vertical direction. The resulting structure resembles a fully connected 3D-mesh architecture. The router port in the vertical direction can be implemented using TSVs. So, in a fully connected mesh architecture, where all the routers are vertically connected by using TSVs to the routers above and below them, the hop count and consequently the communication cost will reduce significantly as compared to a 2D NoC-based system with equal same of routers. However, it is not feasible to provide vertical connections to each and every router because of the manufacturing cost and chip area consumed by the TSVs. It is important to note that the TSV process does not scale with CMOS technology. TSV diameters (4 μm) and pitches (8 μm) are two to three orders of magnitude higher than transistor gate lengths (Kim et al. 2009). Therefore, a fully connected 3D-mesh is more of a theoretical topology. To circumvent this problem, different types of partially connected 3D-mesh topologies have been proposed in the literature. In Liu et al. (2011), four adjacent routers share one TSV. Eight adjacent routers can also share one TSV, as reported in Xu et al. (2010). Another vertically-partially-connected 3D-mesh-NoC architecture has been presented in Dubois et al. (2013). In this architecture, number, position and data flow direction of TSVs can be varied from die to die.

The current work uses the vertically-partially-connected 3D-mesh-NoC architecture and furthermore, tile-based NoC design methodology. It is assumed that the data flow in each TSV can be in both directions. A typical structure has been presented in Fig. 3.2. Here, routers and cores are denoted by r and c, respectively. The number, data flow direction and location of TSVs are the design-time constraints. These parameters are chosen by the designer by taking care of technological constraints, system performance and the overall budget of the system. The number of TSVs that can be afforded in the system depends on the area constraint. Moreover, in Liu et al. (2011) the authors have suggested that it rarely happens that the adjacent routers use their vertical links at the same time. Based on this observation, the work (Liu et al. 2011) suggested that 25% vertical connection between two layers can be a good choice to design a 3D NoC-based system. Hence, this work restricted the vertical links to only 25% of the routers per layer. This number can vary based on the system's requirements such as performance and cost of the system. Furthermore, a spread out arrangement of TSVs that produces the lowest communication cost compared to any other architecture with the same number of TSVs has been reported in Xu et al. (2011). This has motivated the present work to place the TSVs in a spread out fashion, including a constraint on the given number of the TSVs. No two TSVs can be within one-hop distance of each other. This implies that a minimum of two-

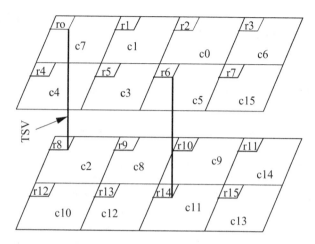

Fig. 3.2 A two layers partially connected 3D NoC-based systems

hop distance should be maintained between the routers having TSVs while placing them. Such distance can vary based on the requirement. In such a scenario, the traditional *dimension order routing* may not be applicable. For vertically-partially-connected 3D-mesh-based-NoC architecture, a deadlock-free *elevator first* routing algorithm has been proposed in Lee and Choi (2013) that has been adopted in present work.

2 TSV Placement and Application Mapping Strategy

The proposed strategy works in two phases. In the first phase, it is assumed that all routers in the NoC have a vertical connection using the TSVs. The application mapping algorithm is run on this topology to produce a task-to-core association. After this, individual TSVs are inspected for traffic flow through them. If the design constraint specifies that only $t\%$ routers in each layer can be 3D-type, we need to modify the remaining routers to 2D. For this, top $t\%$ 3D routers of a layer are preserved which carry the maximum traffic. Moreover, these TSV locations are replicated in successive layers. This strategy also cares to ensure that no two neighbouring routers are 3D in nature. The mapping problem is now solved again on this modified 3D topology to arrive at the final solution. Thus, the mapping algorithm described next works with predefined TSV positions and maps the tasks into the cores of the systems. The application is represented in the form of a task graph with nodes corresponding to the participating tasks and edges identifying the communication requirements between them (in terms of bits per seconds). In the topology graph, nodes correspond to the cores and edges to the links in the topology.

Algorithm 3.1 KL_Map_and_TSV_place

Input: Task graph C and Topology graph T
Output: Mapping of C to T along with position of TSVs in T satisfying restrictions on TSV
    numbers and places.

1: TSV_position ← All routers in T
2: KL_Partition(G, $Task\_List$)
3: Map_clusters($Task\_List$, T)
4: Improvement_Phase(G,T)
5: Sort TSVs in decreasing order of usage
6: Update TSV_positions to hold only top 25% TSV positions giving topology T'
7: Map_clusters($Task\_List$, T')
8: Improvement_Phase(C, T')
9: **return** Mapping

At first, KL_Map_and_TSV_place algorithm partitions the task graph using KL_partitioning procedure and produces a number of clusters each having only two tasks. Now, it maps each such cluster to the topology T. To arrive at a good solution, it tries local changes by applying *swapping* and *flipping* operations in procedure Improvement_Phase. Next, it analyzes the traffic flow through each TSV and sorts the TSVs on descending order of traffic flow. Now, it preserves the top 25% TSVs in the individual layers of the topology. Care is also taken so that no two TSVs are placed within a one-hop distance of each other. The modified topology is noted as T' and the clusters that are coming from KL-partition are mapped onto this modified topology. To achieve a better mapping solution on this modified topology, we once again run the iterative improvement phase and get the final solution.

3 Partitioning Algorithm

Kernighan–Lin partitioning strategy (Kernighan and Lin 1970), originally developed for VLSI physical design, bipartitions a set of modules so that highly connected modules are kept in one partition. In NoC-based system, the strategy uses extended Kernighan–Lin (KL) bi-partitioning, proposed for 2D NoC (Sahu et al. 2014a) design, to identify the closeness of tasks by analyzing their bandwidth requirements. This bi-partitioning is applied (recursively) until only the closest two tasks are left in any of the final partitions. During the mapping phase, these partitions are taken into consideration to minimize the communication cost between mapped tasks. This chapter extends it to design a 3D NoC-based system, integrating the TSV placement problem with it. TSVs are placed after detailed consultation with the application. As a result, it evolves as a complete solution for the 3D NoC-based systems design problem and also incorporates several improvement operations on the initial mapping solution. Algorithm 3.2 has been used to recursively partition the task graph.

Algorithm 3.2 KL-partitioning

 1: **procedure** KL-PARTITIONING(G, l, C)

Input: Task graph $G(C, E)$, l : level of partitioning, $partitioned\_list$: the subset of tasks in C to be partitioned

Output: A list of cluster; $cluster_i$ is the ith level collection of disjoint sets of tasks, initialized to ϕ

 2:    $Cluster_l \leftarrow Cluster_l \bigcup Partitioned\_list$

 3:    **if** $Partitioned\_list$ contains only 2 tasks **then**

 4:        **return**

 5:    **end if**

 6:    $(p_1, p_2) \leftarrow$ Random partitioning of tasks in $Partitioned\_list$

 7:    $KL(G, p_1, p_2)$

 8:    $KL\text{-}partitioning(G, l + 1, p_1)$

 9:    $KL\text{-}partitioning(G, l + 1, p_2)$

10: **end procedure**

 1: **procedure** KL(G, p_1, p_2)

 2:    $present\_partition \leftarrow best\_partition \leftarrow (p_1, p_2)$

 3:    Unlock all tasks

 4:    **while** unlock_task_exist($present\_partition$) **do**

 5:        swap $\leftarrow$ select_next_move($present\_partition$)

 6:        $best\_partition \leftarrow$ get_better_partition(best_partition, present_partition)

 7:        **if** not(cost_fct($best\_partition$) < cost_fct(p_1, p_2)) **then**

 8:            **return** (p_1, p_2) // Terminate, no improvement

 9:        **else**// do another iteration

10:            $(p_1, p_2) \leftarrow best\_partition$

11:            Unlock all tasks

12:        **end if**

13:    **end while**

14: **end procedure**

 1: **procedure** SELECT_NEXT_MOVE(P)

 2:    **for** each unlocked $(c_i \epsilon p_1, c_j \epsilon p_2)$ **do**

 3:        Append costlog.Cost_Fct(swap(P, c_i, c_j))

 4:    **end for**

 5:    **return** (c_i, c_j swap in cost log with lowest cost)

 6: **end procedure**

Initially, at *level*-0, all tasks are in one partition. At *level*-1, there are two partition sets, having partition numbers 0 and 1, each containing half of the nodes of the task graph. At next level (*level*-2), four partitions are generated (two from Pid-0 and two from Pid-1) having partition numbers 0, 1, 2 and 3. This continues until there are only two tasks left in each partition for Mesh. As KL-partitioning results depend on the initial partitioning, we run the algorithm for τ (preset) times, each time starting with a different randomly generated initial partition. The best one is used for subsequent mapping and iterative improvement phase.

In KL, bi-partitioning has been done in a balanced way. Thus, this work considers the number of nodes in a task graph to be an exact power of 2. Otherwise, it incorporates a number of dummy-tasks into the task graph. Dummy-tasks are connected with all tasks including themselves. An edge with cost *zero* connects

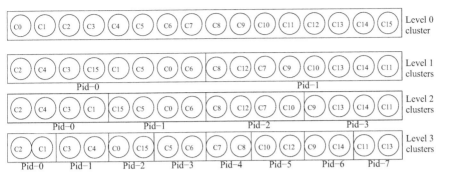

Fig. 3.3 Partition of all levels of a task graph contains 16 tasks

a real task to a dummy task. Edges between dummy tasks are assigned cost *infinity*. The dummy tasks are removed at the end of the mapping phase.

The algorithm takes a task graph (G), level (l) and the set of tasks to be partitioned in *Partitioned_list* as inputs. It is invoked with $l = 0$. It starts partitioning and computes sets of clusters at different levels. In the beginning, it forms a single cluster, cluster[0], which is a single set consisting of all tasks in C. In the next level, it is partitioned into two sets of equal sizes. Cost of the partition is taken to be equal to the sum of edge costs (in the task graph) of tasks belonging to two sides of the partition. Thus, cluster[1] is a collection of two sets. The tasks belonging to a set should be mapped on NoC in the closest proximity of each other, compared to two tasks belonging to two different sets. In general, a set in cluster[k] gives rise to two $(k + 1)$ level disjoint sets in cluster[$k + 1$]. This process continues until the individual sets in a cluster hold only 2 tasks.

The following is the illustration of working of the algorithm. Figure 3.3 demonstrates the algorithm steps. Here, the application contains 16 tasks. As a result, no dummy nodes are added to the task graph. KL-partitioning algorithm is applied upon the task graph and its output is shown in Fig. 3.3. The algorithm recursively bipartitions the task graph until there are 2-tasks left in each partition. At every level, a partition ID (Pid) is assigned to each task based on partitions suggested by the KL-partitioning algorithm. For example, at level-1, partitions with Pid-0 and 1 are formed. At level-2, partitions with Pid-0 and 1 are formed from Pid-0 at level-1 and partitions with Pid-2 and 3 are formed from Partition ID-1 at level-1. The same process is followed at the third level, and so on. In the end, partitions with 2-tasks are obtained which are having the most connectivity between them.

4 Application Mapping onto Mesh-Based 3D NoC Systems

The next task is to assign each of the 2-core graphs to nearby routers of the vertically-partially-connected 3D mesh NoC using *Map_clusters*(). Though, it is possible to attach the 2-core groups to nearby routers arbitrarily, it is still possible to explore optimization that can be achieved by swapping and flipping of the groups in a hierarchical fashion. At each level of partitioning, each core is assigned a partition number. This numbering has been utilized in the address assignment process (initial mapping). Then, an iterative improvement phase has been applied on the initial mapping solution.

4.1   Initial Mapping Phase

In this phase, the final partitions generated using KL-partitioning, are mapped to the respective partition IDs of the mesh-based 3D NoC systems using $Map\_clusters(Task\_List, T)$, as shown in Fig. 3.4. For example, task pair (c_1, c_2) is mapped to the core pair (r_0, r_1). To achieve better solution, iterative improvement is applied on this initial mapping.

4.2   Iterative Improvement Phase

At the end of the partitioning and initial mapping phase, a number of clusters at different levels have been created on the mesh topology. Let, l be the total number of levels that have been created. In the following, the Improvement_Phase algorithm has been presented that chooses two partitions (say, M and N) and attempts various

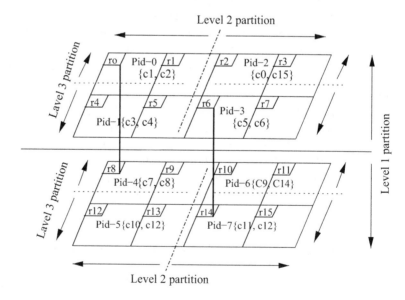

Fig. 3.4 Partition within mesh topology

local changes. Partition M and N are chosen in such a way that they have common parent and level. At first, the tasks of partition N are flipped, keeping M unchanged. The tasks are flipped first along the horizontal axis, then along the vertical axis and finally again along the horizontal axis, creating three new mapping candidates (a total of 4 mapping candidates). The tasks in partition M are now swapped along the horizontal axis. The flipping of partition N is repeated generating four more mappings. In M-partition, swapping is done two more times, first along vertical and then along horizontal axes. Each such case generates 4 new mapping candidates via flipping of tasks in partition N. Therefore, the entire process creates 16 mapping candidates among tasks in partitions M and N. The mapping with the least local communication cost (computed by considering cores in M and N only) is taken. The strategy is then repeated on partition M and N but considering improvement in the global cost for the entire network. The process is continued with the next pair of partitions. If all the partitions in the current level have been completed, the algorithm considers next lower level clusters, until it reaches the level-0 cluster.

```
1:  procedure IMPROVEMENT_PHASE(G, T)
2:      for each level starting from higher level do
3:          for each unlocked partition at this level do
4:              Choose two partitions (say M and N) which have common parent
5:              repeat
6:                  Now, perform flipping operation on tasks of partition N in the
                    order such as horizontal, vertical and horizontal with respect to
                    tasks in partition M and note the local communication cost
7:                  Swap the tasks in partition M horizontally and repeat step 5
8:                  Swap tasks modified M vertically and repeat step 5
9:                  Swap tasks in modified M horizontal and repeat step 5
10:                 The modified mapping of M and N is identified as M' and N'
                    which gives minimum communication cost
11:                 M = M' and N = N'
12:             until global cost minimization
13:             Lock the partition M and N
14:         end for
15:     end for
16: end procedure
```

Let us consider an application mapping in which Improvement_Phase starts with level-3, followed by level-2 and 1. Figure 3.5 shows level-3 operations on partitions M and N. The flip operation is performed on partition N. As there is only a single row in each partition, further flips within them will not result in any better cost. The final mapping after all iterations in level-3 has been shown in Fig. 3.5. Similar flipping is performed at other levels as well which are shown in Figs. 3.6 and 3.7. The final mapping solution has been shown in Fig. 3.2.

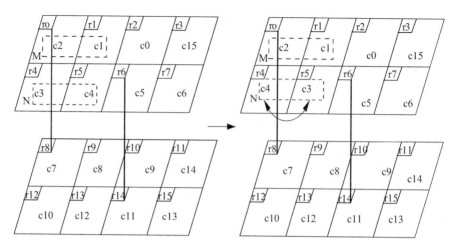

Fig. 3.5 Improvement operation at level-3 partition

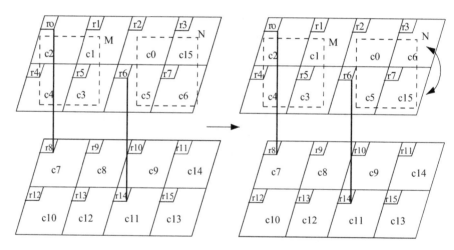

Fig. 3.6 Improvement operation at level-2 partition

5   Experimental Results and Analysis

This section describes the experimental results obtained for a set of bench-mark of NoC systems and compares them with some existing approaches. We use some of the standard benchmarks like PIP, 263ENC-MP3DEC, MWD, MPEG4, VOPD and DVOPD (Sahu and Chattopadyay 2013; Murali and Micheli 2004) and generated some large 64-task and 128-task applications using the task graph generation tool: TGFF (Sahu and Chattopadyay 2013; Sahu et al. 2014a). These benchmarks are noted as G17-G22 and G25-G29 (Sahu et al. 2014a) in Table 3.1. The TGFF parameters used for their generation are as follows: The

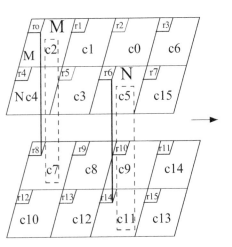 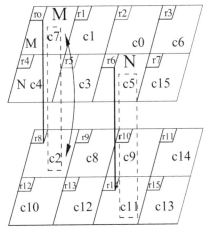

Fig. 3.7 Improvement operation at level-1 partition

Table 3.1 Benchmark applications and their mesh sizes

Benchmarks	No. of tasks	3D Mesh dimensions
PIP	8	$2 \times 2 \times 2, 1 \times 2 \times 4$
263ENC-MP3DEC	12	$2 \times 3 \times 2, 1 \times 3 \times 4$
MWD	12	$2 \times 3 \times 2, 1 \times 3 \times 4$
MPEG4	12	$2 \times 3 \times 2, 1 \times 3 \times 4$
VOPD	16	$2 \times 4 \times 2, 2 \times 2 \times 4$
DVOPD	32	$4 \times 4 \times 2, 2 \times 4 \times 4$
G17	64	$4 \times 8 \times 2, 4 \times 4 \times 4$
G18	64	$4 \times 8 \times 2, 4 \times 4 \times 4$
G19	64	$4 \times 8 \times 2, 4 \times 4 \times 4$
G20	64	$4 \times 8 \times 2, 4 \times 4 \times 4$
G21	64	$4 \times 8 \times 2, 4 \times 4 \times 4$
G22	64	$4 \times 8 \times 2, 4 \times 4 \times 4$
G25	128	$8 \times 8 \times 2, 4 \times 8 \times 4$
G26	128	$8 \times 8 \times 2, 4 \times 8 \times 4$
G27	128	$8 \times 8 \times 2, 4 \times 8 \times 4$
G28	128	$8 \times 8 \times 2, 4 \times 8 \times 4$
G29	128	$8 \times 8 \times 2, 4 \times 8 \times 4$

bandwidths are varied from 50 MB/s to 150 MB/s for some graphs and 10 MB/s to 1500 MB/s for others. For generating both high and low communication graphs, in-out degrees of nodes are varied from 1 to 8. The number of start nodes is also varied to generate different graphs and to see the effect of mapping and TSV placement solution upon them. The bandwidth values for the edges are also generated randomly to get heterogeneous communication behaviour of tasks. The graphs are mapped onto 3D-mesh-NoC-based systems of 2 and 4 layers, as noted in Table 3.1. While generating the mapping solutions, the static performance of mapping has been

Table 3.2 Noxim settings

Parameters	Values
Buffer depth	6
Minimum and maximum packet size	64 flits (32 bits per flit)
Routing	Dimension ordered (xy)
Selection logic	Random
Warmup time	10,000 clk cycles
Simulation time	200,000 clk cycles
Traffic	Table based

evaluated via the communication cost metric. Communication cost is measured as the product of the bandwidth requirement of a pair of tasks and hop count between corresponding mapped tasks, summed over the edges of the application task graph. Dynamic performance (in terms of throughput, latency and energy consumption) of mapping solutions have been obtained via the 3D NoC-based simulator, *Noxim* (Vincenzo et al. 2016), a cycle-accurate simulator. Here, cores send messages by following the traffic pattern in accordance with the edge weights of the task graph. However, the time distribution of the messages follows self-similar nature (Sahu et al. 2014b; Varatkar and Marculescu 2004; Chang and Chen 2008). We have used a similar technique to generate traffic for our applications. Each core generates traffic in a self-similar fashion by aggregating a large number of ON-OFF message sources following Pareto distribution with Hurst parameter, $H = 0.75$, Shape parameters $\alpha_{ON} = 1.5$ and $\alpha_{OFF} = 1.17$ (Kundu et al. 2012). The configuration of the Noxim simulator has been presented in Table 3.2. For design good NoC-based systems, it is expected that the throughput of the network be high, while the average latency be low (Feero and Pande 2009).

5.1   Results on Different TSV Distributions and Mapping Strategies

This section presents experimental results to highlight the efficiency of the proposed mapping strategy KL_Map_and_TSV_place, noted in Sect. 2. As noted earlier, the algorithm takes as input the TSV positions and generates a mapping optimizing the communication cost. Table 3.3 notes the corresponding results. This work has considered four different types of TSV distributions. The experimental results marked *Fully* assume all routers to be 3D in nature, each having a TSV. The columns marked *Symm.* assume a uniform distribution of 25% TSVs, whereas, *Rand.* corresponds to a random distribution of 25% TSVs with the restriction that no two adjacent routers are having TSVs. The TSVs have been placed in consultation with the application, as mentioned in Sect. 2. The values presented in the columns marked *Intl.*, are the communication cost of application, remapped onto the partially

Table 3.3 Communication cost for different applications with various configurations TSV-locations

Layers	Benchmarks	PSMAP (Sahu et al. 2011) 100% Fully	Squeezing (Liu et al. 2011) 25% Symm.	NMAP (Murali and Micheli 2004) 100% Fully	NMAP 25% Symm.	NMAP Rand.	NMAP Intl.	Proposed KL_Map_and_TSV_place 100% Fully	Proposed 25% Symm.	Proposed Rand.	Proposed Intl.
Two	PIP	640	768	640	896	896	896	640	768	768	768
	263ENC-MP3DEC	230.21	230.21	230.94	230.21	268	230.21	230.47	230.47	230.4	230.99
	MWD	1335	1248	1240	1368	1380	1255	1220	1350	1402	1252
	MPEG4	3814	3714	3672	3773	4022.5	3672	3631	3850	4006	3706
	VOPD	4135	4157	4199	4281	4642	4279	4167	4189	4189	4189
	DVOPD	10,032	10,307	9914	10,506	10,740	10,404	9618	9784	9838	9726
	G17	42,465.1	50,626	36,800.49	39,387.91	45,362.3	39,387.91	36,567.65	39,651.87	42,876.54	38,576.63
	G18	8046.1	9814.02	7684.29	7713.99	8361.63	7579.85	6700.5	6898.52	7117.9	6847.48
	G19	7852.23	9930.1	8993.43	9212.45	9284.82	8886.84	7309.54	7321.96	7601.37	7231.05
	G20	12,156.34	147,380.8	120,675.94	132,741.59	134,904.02	117,768.57	114,242.09	117,615.06	121,590.9	114,297.94
	G21	111,673.1	129,985.18	105,054.88	111,384.42	124,357.06	111,384.42	102,501.54	108,654.56	113,456.65	116,985.25
	G22	48,300.16	54,354.19	49,722.7	51,802.42	58,848.55	51,509.32	45,378.92	46,040.36	46,821.3	45,223.15
	G25	160,396.69	166,271.15	131,295.61	201,561.16	216,127.15	145,123.25	125,219.16	171,259.12	180,156.17	134,161.25
	G26	20,781.36	29,231.51	18,239.18	20,459.38	21,364.35	19,201.61	18,154.39	21,756.12	22,456.32	20,613.78
	G27	74,391.23	75,128.27	51,929.12	58,251.15	68,152.51	53,489.71	49,110.13	76,210.15	77,516.12	52,374.61
	G28	460,148.19	551,751.31	427,896.23	494,782.15	501,894.85	442,131.51	402,789.74	451,238.34	584,592.58	432,371.81
	G29	381,928.13	396,121.45	291,247.32	371,278.64	404,527.12	312,261.48	291,782.38	367,458.15	421,856.15	305,166.55
	Rank	1.07	1.26	1.05	1.18	1.26	1.10	1	1.12	1.18	1.05

(continued)

Table 3.3 (continued)

Layers	Benchmarks	PSMAP (Sahu et al. 2011) 100% Fully	Squeezing (Liu et al. 2011) 25% Symm.	NMAP (Murali and Micheli 2004) 100% Fully	25% Symm.	25% Rand.	Intl.	Proposed KL_Map_and_TSV_place 100% Fully	25% Symm.	25% Rand.	Intl.
Four	PIP	640	896	640	896	896	896	640	896	896	896
	263ENC-MP3DEC	230.43	230.51	230.47	230.47	268.47	230.51	230.51	230.51	230.51	230.51
	MWD	1216	2015	1723	2107	2245	1965	1332	2110	2345	1845
	MPEG4	3854	4225	3991.78	4291.86	4681.37	3940.21	3712.95	4387.16	4689.37	3960.12
	VOPD	4513	14,415	4652.18	15,073.18	5319.16	4900	4317	5105.74	5479.16	4773
	DVOPD	12,076	11,257	16,548	17,389	19,789	11,668	9812	19,186	20,789	10,004
	G17	47,305.92	54,251.15	51,291.91	57,166.16	59,137.92	52,507.96	43,128.41	56,912.16	59,736.92	45,622.87
	G18	8513.02	9912.07	8217.17	8753.18	8924.74	8459.07	7875.83	8614.91	8827.74	6847.07
	G19	8943.63	9985.12	8998.84	9318.17	9617.56	8984.38	8317.97	9418.8	9657.56	7266.12
	G20	16,875.65	15,345	15,864	20,169	24,173	18,891.95	18,967.85	19,813	24,975	17,726.21
	G21	17,854	13,172.01	15,327.16	17,569.18	20,897.19	15,931.9	10,768.18	16,891.13	22,867.19	15,701.13
	G22	52,871.97	55,129.03	50,169.63	62,199.67	65,289.18	58,671.28	48,191.73	61,253.85	63,284.18	51,989.62
	G25	20,357.69	173,242.6	147,844.18	226,108.17	296,781.71	152,145.6	138,727.82	176,538.18	197,741.71	149,812.1
	G26	27,431.8	30,124.17	20,010.7	27,907.87	30,876.61	22,972.12	17,768.19	20,187.91	23,836.61	18,321.26
	G27	79,732.18	77,451.6	55,781.87	62,197.19	65,781.02	57,361.02	50,189.82	56,278.76	58,771.02	53,241.18
	G28	49,678.96	56,715.31	43,784.67	55,189.21	58,163.19	47,174.12	40,186.21	45,089.18	51,169.19	42,981.12
	G29	491,247.81	423,506.18	323,472.98	431,872.81	456,193.17	358,215.72	281,927.76	321,926.18	353,145.17	2,931,583.12
	Rank	**1.14**	**1.40**	**1.13**	**1.49**	**1.48**	**1.19**	**1**	**1.27**	**1.39**	**1.09**

Bold values indicate the entire results presented in this table

Table 3.4 Time comparison between PSMAP and the proposed approach

Benchmarks	CPU time in sec	
	PSMAP (Sahu et al. 2011)	Proposed KL_Map_and_TSV_place
G17	139.25	40.45
G21	180.88	58.67
G25	257.74	90.47

connected 3D-mesh network. The results have been compared with PSMAP (Sahu et al. 2011), Squeezing (Liu et al. 2011) and NMAP (Murali and Micheli 2004). Out of these three, NMAP is a constructive heuristic algorithm, whereas PSMAP is a particle swarm optimization based approach. We have extended NMAP and PSMAP, originally proposed for 2D-NoC, to 3D. PSMAP assumes a fully connected 3D NoC-based systems. From Table 3.3, it can be noted that compared to the KL_Map_and_TSV_place fully connected version, PSMAP and NMAP show 7% and 5%, on an average, increase in communication cost for two layers, whereas, for four layers, 9% and 10% increase in communication cost could be observed. However, compared to KL_Map_and_TSV_place fully connected version, works Squeezing (Liu et al. 2011), NMAP (*Intl.*) and KL_Map_and_TSV_place (*Intl.*) are about 26%, 10% and 5% inferior, on an average for two layers, whereas, for four layers such degradations are 35%, 10% and 5%, respectively. This shows the merit of the proposed mapping together with TSV placement approach. Table 3.4 compares the CPU time requirements of PSMAP and the proposed approach for a number of benchmarks. On average, our approach requires one-third the time needed for PSMAP.

5.2 Impact of TSV Position Selection

Table 3.5 enumerates the experimental results of integrated TSV position selection and mapping. The column marked *Fully* corresponds to the situation in which all routers are 3D in nature. As suggested in Sect. 2, we next keep only 25% highly utilized TSVs. Communication cost values are recomputed such that interlayer message flow uses elevator-first algorithm (Dubois et al. 2013). Naturally, communication cost degrades. Next, it performs a remapping of tasks with the current TSV distribution. Compared to the fully connected configuration, the 25% TSV case without remapping increases communication cost by 31% and 84%, on an average for two and four layers, respectively. A remapping of tasks, now, creates a final solution with communication cost degradation restricted to 5% with respect to the fully connected version for two and four layers.

Table 3.5 Communication cost before and after re-mapping with of 25% highly utilized TSVs retained

	Two-layer			Four-layer		
Benchmarks	Fully	25% TSVs	Re-map with 25% TSVs	Fully	25% TSVs	Re-map with 25% TSVs
PIP	640	968	768	640	1296	896
263ENC-MP3DEC	230.47	430.99	230.99	230.51	530.51	230.51
MWD	1220	1852	1252	1332	3745	1845
MPEG4	3631	4806	3706	3712.95	4080.12	3960.12
VOPD	4167	4789	4189	4317	4893	4773
DVOPD	9618	9926	9726	9812	10,904	10,004
G17	36,567.65	48,676.63	38,576.63	43,128.41	45,799.87	45,622.87
G18	6700.5	8947.48	6847.48	7875.83	7997.07	6847.07
G19	7309.54	8331.05	7231.05	8317.97	7996.12	7266.12
G20	114,242.09	164,397.94	114,297.94	18,967.85	20,826.21	17,726.21
G21	102,501.54	177,085.25	116,985.25	10,768.18	23,801.13	15,701.13
G22	45,378.92	46,323.15	45,223.15	48,191.73	59,089.62	51,989.62
G25	125,219.16	154,261.25	134,161.25	138,727.82	199,912.1	149,812.1
G26	18,154.39	22,713.78	20,613.78	17,768.19	19,921.26	18,321.26
G27	49,110.13	55,474.61	52,374.61	50,189.82	58,741.18	53,241.18
G28	402,789.74	462,471.81	432,371.81	40,186.21	49,091.12	42,981.12
G29	291,782.38	325,266.55	305,166.55	281,927.76	2,991,683.12	293,158.12
Rank	**1.00**	**1.31**	**1.05**	**1.00**	**1.98**	**1.09**

Bold values indicate the entire results presented in this table

5.3 Dynamic Performance of Different Mapping and TSV Configurations

To measure the proficiency of individual mapping techniques together with different TSV configuration in partially-connected 3D-mesh-NoC-based systems, simulation has been performed using Noxim-3D simulator (Vincenzo et al. 2016). We have incorporated the Elevator-first (Bahmani et al. 2012; Dubois et al. 2013) routing algorithm into the Noxim. The TSV positions, generated from the proposed methodology, have been provided to Noxim. Synthetic self-similar traffic has been generated, by obeying the communication requirement of tasks in the application. Self-similar traffic has been observed in typical video and networking applications (Varatkar and Marculescu 2004). Tables 3.6 and 3.7 show the results of throughput, latency and average packet energy (μJ) for the benchmarks, considering two and four layers for 3D NoC. As it can be noted from the table, fully connected 3D NoC configuration gives the best performance in terms of all the three factors. Using 25% TSVs results in degradation of all the three parameters. However, intelligent TSV placement with task remapping improves the solution quality compared to 25% *Symm.* and the *25% Rand.* in throughput, latency and packet energy.

Table 3.6 Comparison of throughput, latency and energy (μJ) of different mapping and TSV placement strategies with two layers 3D NoCs

Two layers

Parameters 2-17 — Proposed KL_Map_and_TSV_place with

	Benchmarks: 263ENC-MP3DEC				Benchmarks: MWD				Benchmarks: MPEG4				Benchmarks: VOPD			
	100%	25%			100%	25%			100%	25%			100%	25%		
	Fully	Symm.	Rand.	Intl.	Fully	Symm.	Rand.	Intl.	Fully	Symm.	Rand.	Intl.	Fully	Symm.	Rand.	Intl.
Throughput	0.80	0.80	0.80	0.80	0.72	0.69	0.66	0.73	0.58	0.52	0.45	0.56	0.68	0.63	0.06	0.66
Latency	77,461	77,461	77,461	77,461	99,263	99,871	99,951	98,123	98,126	99,625	99,912	97,216	95,129	98,121	99,726	95,012
Pkt. energy	9.73	9.73	9.73	9.73	12.35	13.12	13.98	12.11	13.51	13.69	14.25	13.12	13.15	13.98	14.25	13.25

	Benchmarks: DVOPD				Benchmarks: G17				Benchmarks: G18				Benchmarks: G19			
	100%	25%			100%	25%			100%	25%			100%	25%		
	Fully	Symm.	Rand.	Intl.	Fully	Symm.	Rand.	Intl.	Fully	Symm.	Rand.	Intl.	Fully	Symm.	Rand.	Intl.
Throughput	0.59	0.54	0.51	0.58	0.52	0.48	0.43	0.50	0.65	0.61	0.60	0.64	0.63	0.58	0.53	0.61
Latency	97,516	98,871	99,659	97,210	99,105	100,121	101,521	98,562	99,850	100,215	101,259	99,520	99,821	101,250	107,259	98,125
Pkt. energy	12.19	13.01	13.25	12.01	14.12	14.44	14.89	14.01	12.01	12.39	12.97	12.21	13.19	13.58	13.89	13.01

	Benchmarks: G20				Benchmarks: G21				Benchmarks: G22				Benchmarks: G25			
	100%	25%			100%	25%			100%	25%			100%	25%		
	Fully	Symm.	Rand.	Intl.	Fully	Symm.	Rand.	Intl.	Fully	Symm.	Rand.	Intl.	Fully	Symm.	Rand.	Intl.
Throughput	0.68	0.63	0.60	0.67	0.72	0.68	0.65	0.75	0.52	0.48	0.46	0.51	0.70	0.64	0.62	0.69
Latency	99,785	102,315	109,135	98,732	107,238	108,125	121,321	106,521	99,954	105,213	108,761	98,725	97,351	98,912	99,816	96,129
Pkt. energy	14.01	14.85	14.97	14.25	14.24	14.81	14.92	14.43	14.38	14.59	14.87	14.51	11.85	11.93	11.98	11.89

	Benchmarks: G26				Benchmarks: G27				Benchmarks: G28				Benchmarks: G29			
	100%	25%			100%	25%			100%	25%			100%	25%		
	Fully	Symm.	Rand.	Intl.	Fully	Symm.	Rand.	Intl.	Fully	Symm.	Rand.	Intl.	Fully	Symm.	Rand.	Intl.
Throughput	0.68	0.63	0.60	0.66	0.55	0.52	0.48	0.54	0.60	0.53	0.51	0.58	0.65	0.60	0.53	0.63
Latency	98,125	98,853	99,851	98,029	98,341	98,921	99,725	97,359	89,251	95,321	98,351	87,215	95,341	96,232	98,921	95,031
Pkt. energy	12.35	12.84	12.98	12.39	14.93	15.05	15.28	14.99	13.25	13.65	13.82	13.42	14.25	14.59	14.63	14.42

Table 3.7 Comparison of throughput, latency and energy (μJ) of different mapping and TSV placement strategies with four layers 3D NoCs

Four layers

Parameters	Proposed KL_Map_and_TSV_place with															
	Benchmarks: 263ENC-MP3DEC				Benchmarks: MWD				Benchmarks: MPEG4				Benchmarks: VOPD			
	100%	25%			100%	25%			100%	25%			100%	25%		
	Fully	Symm.	Rand.	Intl.	Fully	Symm.	Rand.	Intl.	Fully	Symm.	Rand.	Intl.	Fully	Symm.	Rand.	Intl.
Throughput	0.78	0.78	0.78	0.78	0.68	0.63	0.60	0.67	0.55	0.50	0.48	0.54	0.62	0.56	0.52	0.61
Latency	78,641	78,641	78,641	78,641	99,921	101,256	106,239	98,956	99,251	102,512	99,812	98,125	96,219	97,321	98,561	95,213
Pkt. energy	9.73	9.73	9.73	9.73	11.31	11.85	11.98	11.35	13.01	13.26	13.59	13.21	12.48	12.68	12.81	12.52

Parameters	Benchmarks: DVOPD				Benchmarks: G17				Benchmarks: G18				Benchmarks: G19			
	100%	25%			100%	25%			100%	25%			100%	25%		
	Fully	Symm.	Rand.	Intl.	Fully	Symm.	Rand.	Intl.	Fully	Symm.	Rand.	Intl.	Fully	Symm.	Rand.	Intl.
Throughput	0.51	0.48	0.43	0.50	0.47	0.42	0.40	0.45	0.61	0.58	0.56	0.60	0.58	0.55	0.52	0.57
Latency	98,216	98,791	99,562	97,321	98,956	99,529	99,821	96,219	98,621	99,721	99,958	97,028	98,516	98,968	99,973	96,213
Pkt. energy	11.89	11.92	12.01	11.90	13.89	13.92	14.01	13.90	11.28	11.58	11.98	11.32	11.87	11.91	11.99	11.89

Parameters	Benchmarks: G20				Benchmarks: G21				Benchmarks: G22				Benchmarks: G25			
	100%	25%			100%	25%			100%	25%			100%	25%		
	Fully	Symm.	Rand.	Intl.	Fully	Symm.	Rand.	Intl.	Fully	Symm.	Rand.	Intl.	Fully	Symm.	Rand.	Intl.
Throughput	0.62	0.56	0.53	0.61	0.67	0.63	0.60	0.66	0.48	0.43	0.40	0.47	0.65	0.61	0.58	0.54
Latency	98,985	99,629	99,831	97,521	109,231	109,925	112,396	107,512	106,915	108,712	109,615	105,135	98,635	98,891	99,876	97,321
Pkt. energy	13.51	13.89	13.97	13.55	13.58	13.85	13.99	13.59	13.59	13.79	13.89	13.62	10.23	10.56	10.89	10.35

Parameters	Benchmarks: G26				Benchmarks: G27				Benchmarks: G28				Benchmarks: G29			
	100%	25%			100%	25%			100%	25%			100%	25%		
	Fully	Symm.	Rand.	Intl.	Fully	Symm.	Rand.	Intl.	Fully	Symm.	Rand.	Intl.	Fully	Symm.	Rand.	Intl.
Throughput	0.63	0.60	0.58	0.61	0.48	0.44	0.41	0.47	0.55	0.51	0.46	0.55	0.62	0.58	0.56	0.51
Latency	98,815	98,961	99,739	97,351	98,735	98,862	99,531	96,921	97,652	98,321	99,681	96,812	97,321	96,232	99,685	96,125
Pkt. energy	11.87	11.92	11.99	11.89	13.56	13.85	13.97	13.58	12.58	12.78	12.96	12.61	13.59	13.76	13.91	13.61

6 Conclusion

In this chapter, we have presented a strategy to design 3D-mesh-based NoC with restricted vertical interconnects via TSVs. An application mapping policy developed around Kernighan–Lin bi-partitioning approach has been designed and integrated with the TSV position selection. In these techniques, the KL-partitioning strategy has been used to ensure that the tasks that communicate with each other frequently are within the same partition. This is required for our initial mapping phase. Our proposed improvement technique is then applied upon the initial mapping to arrive at better solutions. The communication cost of our mapping solutions has been compared with existing mapping approaches. It shows good improvement in both solution quality and execution time for most of the applications. Comparison of average network latency and energy consumption has also been carried out. It may be noted that the strategies based on iterative improvement policy depend heavily on the initial solution. A poor initial solution will make the search process to explore only the nearby cases, restricting the solution quality. All these motivate us to develop a constructive heuristic for 3D NoC-based application mapping and TSV placement problem. Our next chapter presents a constructive heuristic for TSV placement together with application mapping onto 3D NoC.

References

Bahmani, M., Sheibanyrad, A., Petrot, F., Dubois, F., & Durante, P. (2012). A 3D-NoC router implementation exploiting vertically-partially-connected topologies. In *Proceedings of Computer Society Annual Symposium on VLSI (ISVLSI)* (pp. 9–14).

Chang, K. C., & Chen, T. F. (2008). Low-power algorithm for automatic topology generation for application-specific networks-on-chips. *IET Computers and Digital Techniques, 2*(3), 239–249.

Dubois, F., Sheibanyrad, A., Petrot, F., & Bahmani, M. (2013). Elevator-first: A deadlock-free distributed routing algorithm for vertically partially connected 3D-NoCs. *IEEE Transactions on Computers, 62*(3), 609–615.

Feero, B. S., & Pande, P. P. (2009). Networks-on-Chip in a three dimensional environment: A performance evaluation. *IEEE Transactions on Computers, 58*(1), 32–45.

Kernighan, B., & Lin, S. (1970). An efficient heuristic procedure for partitioning graphs. *Bell System Technical Journal, 49*(2), 291–307.

Kim, D. H., Athikulwongse, K., & Lim, S. K. (2009) A study of through-silicon-via impact on the 3D stacked IC layout. In *Proceedings of International Conference on Computer-Aided Design (ICCAD)* (pp. 674–680).

Kundu, S., Soumya, J., & Chattopadhyay, S. (2012). Design and evaluation of mesh-of-tree based network-on-chip using virtual channel router. *Microprocessors and Microsystems, 36*(6), 471–488.

Lee, J., & Choi, K. (2013). A deadlock-free routing algorithm requiring no virtual channel on 3D-NoCs with partial vertical connections. In *Proceedings of International Symposium on Networks on Chip (NoCS)* (pp. 1–2).

Liu, C., Zhang, L., Han, Y., & Li, X. (2011). Vertical interconnects squeezing in symmetric 3D mesh network-on-chip. In *Proceedings of Asia and South Pacific Design Automation Conference (ASP-DAC)* (pp. 357–362).

Murali, S., & De Micheli, G. (2004). Bandwidth constrained mapping of cores onto NoC architectures. In *Proceedings of Design, Automation and Test in Europe (DATE)* (pp. 896–901).

Sahu, P. K., & Chattopadhyay, S. (2013). A survey on application mapping strategies for network-on-chip design. *Journal of System Architecture, 59*(1), 60–76.

Sahu, P. K., Manna, K., Shah, N., Chattopadhyay, S. (2014a). Extending Kernighan–Lin partitioning heuristic for application mapping onto network-on-chip. *Journal of Systems Architecture, 60*, 562–578.

Sahu, P. K., Shah, T., Manna, K., & Chattopadhyay, S. (2014b). Application mapping onto mesh based network-on-chip using discrete particle swarm optimization. *IEEE Transactions on Very Large Scale Integration (VLSI) Systems, 22*(2), 300–312.

Sahu, P. K., Venkatesh, P., Gollapalli, S., & Chattopadhyay, S. (2011). Application mapping onto mesh structured network-on-chip using particle swarm optimization. In *Proceedings of Computer Society Annual Symposium on VLSI (ISVLSI)* (pp. 335–336).

Varatkar, G. V., & Marculescu, R. (2004). On-chip traffic modelling and synthesis for MPEG-2 video applications. *IEEE Transactions on Very Large Scale Integration (VLSI) Systems, 12*(1), 108–119.

Vincenzo, C., Andrea, M., Salvatore, M., Maurizio, P., & Davide, P. (2016). Cycle-accurate network on chip simulation with noxim. *ACM Transactions on Modeling and Computer Simulation, 27*(1), 4:1–4:25.

Xu, T. C., Liljeberg, P., & Tenhunen, H. (2010). A study of through silicon via impact to 3D network-on-chip design. In *Proceedings of International Conference on Electronics and Information Engineering (ICEIE)* (pp. 333–337).

Xu, T. C., Liljeberg, P., & Tenhunen, H. (2011). Optimal number and placement of through silicon vias in 3D network-on-chip. In *Proceedings of International Symposium on Design and Diagnostics of Electronic Circuits & Systems (DDECS)* (pp. 105–110).

Chapter 4
A Constructive Heuristic for Designing a 3D NoC-Based Multi-Core Systems

The design strategy of 3D NoC system proposed in the previous chapter, based on iterative improvement policy, depends on the initial solution to a large extent. A poor initial solution will make the search process to explore only the nearby cases, restricting the solution quality. Though for many of the example cases the iterative mapping is producing results better than NMAP (Murali and Micheli 2004a), it is not true for all the cases. This motivates us to look for constructive solutions to design 3D NoC-based systems that strive to build the solution directly, rather than iterative improvement of an initial solution taken as input. The main idea of the algorithm is to select the best position for the highest communicating task at every point of time during mapping. The position of the next highest communicating task is fixed using a predictive search. The salient features of the approach are as follows:

1. A constructive mapping algorithm based on prediction has been proposed to reduce the communication cost.
2. The communication cost values of the mapping solutions of our approach have been compared with other existing strategies.
3. Comparison of dynamic performance (in terms of average network latency) and energy consumption has also been carried out.

1 Proposed Heuristic for TSV Placement and Application Mapping

In this section, we propose a constructive heuristic based on prediction to map tasks and placing TSVs in the mesh-based 3D NoC. The proposed algorithm has been discussed next. Moreover, it has been explained with an example.

© Springer Nature Switzerland AG 2020
K. Manna, J. Mathew, *Design and Test Strategies for 2D/3D Integration for NoC-based Multicore Architectures*, https://doi.org/10.1007/978-3-030-31310-4_4

1.1   Algorithm Philosophy

First, the edges of the task graph are sorted on descending communication require-ments, specified in edge labels. Let the edge with the maximum bandwidth be $l = (c_1, c_2)$ from task c_1 to task c_2. Mapping process starts with this edge. Also let the total bandwidth requirement of task c_1 be greater than the requirement of c_2. The mapping process generates a solution with task c_1 mapped to each core in the topology. For each placement of task c_1, the remaining tasks are mapped judiciously to obtain a good solution. The mapping with the minimum communication cost is accepted as the final solution.

Algorithm 4.1 Map_and_TSV_place

Require: Task graph C and Topology graph T
Ensure: Mapping of C to T along with position of TSVs in T satisfying restrictions on TSV numbers and places.

1: $TSV\_positions \leftarrow$ All routers in T
2: Map_Task_Graph($C, T, TSV\_position$)
3: Sort TSVs on decreasing order of usage
4: Update $TSV\_positions$ to hold only top 25% TSV positions
5: Map_Task_Graph($C, T, TSV\_positions$)

1: **procedure** $Map\_Task\_Graph(C, T, TSV\_position)$
2:      Sort edges of C on decreasing bandwidth requirement
3:      Best_Cost $\leftarrow \infty$
4:      Best_Mapping $\leftarrow \phi$
5:      Let (c_1, c_2) be the edge with highest bandwidth requirement
6:      Set $c \leftarrow c_1$ if c_1 has higher communication requirement than c_2 else $c \leftarrow c_2$

7:      **for** each unmapped core position r of T **do**
8:          Map c to r
9:          Mark c and r as mapped
10:          **while** there exist unmapped task in C **do**
11:              Find unmapped task d doing maximum communication with mapped tasks
12:              Evaluate mapping costs for all one-hop mappings of d to unmapped cores
13:              Identify core positions R giving same minimum costs
14:              **for** each position p in R **do**
15:                  Map d to p
16:                  Perform mapping of remaining tasks by picking them in decreasing communicati and attaching to any one-hop distance core with minimum cost
17:                  Evaluate mapping
18:                  Map d to the position in R giving minimum cost
19:              **end for**
20:          **end while**
21:          Evaluate mapping

Suppose that the task c_1 is mapped onto the router r_1. Core r_1 has a few neighbours in the topology graph that are one-hop away from it. So, each neighbour can be a potential candidate for mapping task c_2. In general, at any instant of execution of the algorithm, a subset of tasks have been mapped to the cores. Let,

```
22:          if mapping cost < Best_Cost then
23:              Best_Cost ← mapping cost
24:              Best_mapping ← this mapping
25:          end if
26:      end for
27:      return Best_Cost, Best_mapping
28: end procedure
```

C' and R' represent the sets of the already mapped tasks and their associated cores, respectively. The algorithm now considers those edges of the task graph of which exactly one task has already been mapped. It selects such an edge with the highest bandwidth requirement. Let, c_i be the unmapped task of that edge. Now, it tries to place the task c_i at each core located one-hop away from any core in R'. It determines the mapping cost by considering the sub-graph consisting of tasks in the set $C' \bigcup \{c_i\}$. If there exists a unique mapping with minimum cost, that mapping of c_i is considered and the process continues with the next unmapped task selected in a similar fashion. However, for multiple mappings of task c_i all of which have same cost, let $P = \{p_1, p_2, \ldots, p_k\}$ be the set of candidate positions for task c_i resulting in same cost for the sub-graph with vertices $C' \bigcup \{c_i\}$. Among all these k positions, suppose, the algorithm selects p_1, for the time being, to be the mapping of c_i. With this, it proceeds to find an association between remaining cores and the remaining tasks in a similar manner, noted previously. However, the algorithm now does not differentiate between the contending positions with minimum cost value. Instead, it takes the first such position and continues with the mapping of remaining tasks. When all tasks of the task graph have been mapped, the cost of that mapping is taken as the predicted cost of selecting core position p_1 for task c_i. Similarly, for remaining $k - 1$ positions, $p_2, p_3, \ldots, p_k$, costs are evaluated and the task c_i is associated with the core position with the minimum predicted cost. The process continues by selecting the next task.

Thus, for each of the possible mappings of the initial task, the algorithm generates the mapping for all other tasks of the task graph. The mapping resulting in the minimum cost is taken as the final mapping solution. The entire procedure has been detailed in Algorithm 4.1.

For example, PIP contains eight tasks and eight edges. The edge weights are 128, 64, 64, 64, 64, 64, 64 and 64. Figure 4.1 represents the corresponding task graph for PIP. In Map_and_TSV_place, at the first stage, it is assumed that each core in the tier has vertical connection. Now, Map_Task_Graph maps PIP on to $2 \times 2 \times 2$ 3D NoC-based system which has been shown in Fig. 4.2. Here, r_i and c_i represent the cores and tasks, respectively. The maximum weight (128) edge (c_2, c_1) is selected and task c_2 is mapped onto the core r_1 of 3D NoC, as shown in Fig. 4.2. Task c_1 can now be mapped onto three possible cores r_2, r_3 or r_5 the already mapped task c_2. For every core position, the task c_1 will be mapped. At first, say, task c_1 is mapped onto core r_5. Tasks c_3 and c_5 are communicating with already mapped tasks c_1 and c_2 with edge weight of 64. The procedure chooses task c_5 for mapping next.

Fig. 4.1 Task graph for
application PIP

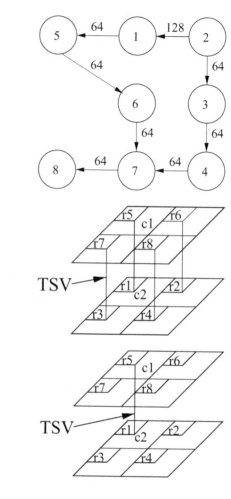

Fig. 4.2 A $2 \times 2 \times 2$ 3D
NoC-based systems with all
router having vertical
connection

Fig. 4.3 A $2 \times 2 \times 2$
partially connected 3D NoC

Similarly, c_5 can be mapped onto any of the four cores r_2, r_3, r_6 or r_7. First, it maps
the task c_5 onto core r_6. Subsequently the task c_5 will be mapped onto the other core
locations as well. The best one with minimum communication cost will be chosen.
In a similar fashion all task will be mapped onto the 3D NoC-based systems and
corresponding mapping cost will be noted. This mapping cost corresponds to the
mapping of task c_2 to core r_1. Next, task c_2 is mapped to core r_2. The entire process
is repeated. Thus, task c_2 will be mapped to every core in the NoC and remaining
tasks will be mapped in a similar fashion. The best mapping will be noted with the
minimum communication cost.

Now, every vertical connection is inspected by considering the best mapping solution. The vertical connections with high traffic flows that honored the
technological constraints are preserved. The remaining vertical connections are
removed from the topology. Such topology has been shown in Fig. 4.3. Thereafter,
Map_and_TSV_place in the second stage applies a similar mapping technique on
this new 3D NoC topology and best mapping is taken as the solution.

2 Experimental Results and Analysis

This section presents the experimental results on different NoC benchmarks and compares these with other existing techniques and the proposed technique in the previous chapter (KL). The simulation environment is the same as in Sect. 5.

2.1 Results on Different TSV Distributions and Mapping Strategies

To start with, we present a few results to highlight the efficiency of our mapping strategy Map_Task_Graph, noted in Sect. 1. Table 4.2 notes the corresponding results. We have experimented with three different types of TSV distributions. The results marked *Fully* assume all routers to be 3D in nature, each having a TSV. The columns marked *Symm.* assume a uniform distribution of TSVs, whereas, *Rand.* corresponds to the random distribution of 25% TSVs with the restriction that no two adjacent routers are having TSVs. From Table 4.2, it can be noted that compared to the Map_Task_Graph fully connected version, PSMAP, NMAP and KL show 14%, 12% and 7% on an average, increase in the communication cost for two layers. For four layers, 19%, 20% and 9% increase in communication cost could be observed. However, compared to Map_Task_Graph fully connected version, works Squeezing (Liu et al. 2011a), NMAP (*Intl.*) and KL (*Intl.*) are about 36%, 17% and 12% inferior, on an average for two layers. For four layers such degradations are 51%, 27% and 15%, respectively. This shows the merit of the proposed mapping together with TSV placement approach. Table 4.1 compares the CPU time requirements of PSMAP and the proposed approach for a number of benchmarks. From the table, it can be noted that PSMAP takes more CPU time than the proposed strategy (Table 4.1).

2.2 Impact of TSV Position Selection

Now, we enumerate the results of integrated TSV position selection and mapping. Table 4.3 notes the corresponding results. The column marked 'Fully' corresponds to the situation in which all routers are 3D in nature. As suggested in Sect. 1, we next

Table 4.1 Execution time comparison between PSMAP and the proposed method

Benchmarks	CPU time in s	
	PSMAP (Sahu et al. 2011)	Proposed Map_and_TSV_place
G17	139.25	73.45
G21	180.88	80.47
G25	257.74	98.67

Table 4.2 Communication cost for different applications with various configurations TSV-locations

Layers	Benchmarks	PSMAP (Sahu et al. 2011)	Squeezing (Liu et al. 2011a)	NMAP (Murali and Micheli 2004a)				KL				Proposed strategy			
		100%	25%	100%	25%			100%	25%			100%	25%		
		Fully	Symm.	Fully	Symm.	Rand.	Intl.	Fully	Symm.	Rand.	Intl.	Fully	Symm.	Rand.	Intl.
Two	PIP	640	768	640	896	896	896	640	768	768	768	640	768	768	768
	263ENC-MP3DEC	230.21	230.21	230.94	230.21	268	230.21	230.47	230.47	230.4	230.99	230.41	230.41	230.41	230.42
	MWD	1335	1248	1240	1368	1380	1255	1220	1350	1402	1252	1216	1260	1320	1248
	MPEG4	3814	3714	3672	3773	4022.5	3672	3631	3850	4006	3706	3600	3773	3773	3632
	VOPD	4135	4157	4199	4281	4642	4279	4167	4189	4189	4189	4119	4119	4135	4119
	DVOPD	10,032	10,307	9914	10,506	10,740	10,404	9618	9784	9838	9726	9544	9656	9672	9592
	G17	42,465.1	50,626	36,800.49	39,387.91	45,362.3	39,387.91	36,567.65	39,651.87	42,876.54	38,576.63	36,295.6	40,445.8	40,554.8	40,198.9
	G18	8046.1	9814.02	7684.29	7713.99	8361.63	7579.85	6700.5	6898.52	7117.9	6847.48	6094.11	6393.71	6398.22	6280.91
	G19	7852.23	9930.1	8993.43	9212.45	9284.82	8886.84	7309.54	7321.96	7601.37	7231.05	6415.87	6767.34	6921.29	6717.12
	G20	12,156.34	147,380.8	120,675.94	132,741.59	134,904.02	117,768.57	114,242.09	117,615.06	121,590.9	114,297.94	102,056	116,279	117,199	110,831
	G21	111,673.1	129,985.18	105,054.88	111,384.42	124,357.06	111,384.42	102,501.54	108,654.56	113,456.65	116,985.25	101,256	112,453	116,523	107,280
	G22	48,300.16	54,354.19	49,722.7	51,802.42	58,848.55	51,509.32	45,378.92	46,040.36	46,821.3	45,223.15	41,198.8	49,241.1	49,359.5	48,271.4
	G25	160,396.69	166,271.15	131,295.61	201,561.16	216,127.15	145,123.25	125,219.16	171,259.12	180,156.17	134,161.25	101,235.32	124,561.23	141,245.32	112,315.11
	G26	20,781.36	29,231.51	18,239.18	20,459.38	21,364.35	19,201.61	18,154.39	21,756.12	22,456.32	20,613.78	13,189.28	15,075.24	16,238.37	14,725.12
	G27	74,391.23	75,128.27	51,929.12	58,251.15	68,152.51	53,489.71	49,110.13	76,210.15	77,516.12	52,374.61	47,129.49	49,782.14	51,349.19	48,238.17
	G28	460,148.19	551,751.31	427,896.23	494,782.15	501,894.85	442,131.51	402,789.74	451,238.34	584,592.58	432,371.81	352,189.12	457,892.37	481,297.23	381,112.11
	G29	381,928.13	396,121.45	291,247.32	371,278.64	404,527.12	312,261.48	291,782.38	367,458.15	421,856.15	305,166.55	349,856.14	360,147.19	421,947.23	242,312.91
	Rank	**1.14**	**1.36**	**1.12**	**1.26**	**1.35**	**1.17**	**1.07**	**1.20**	**1.26**	**1.12**	**1**	**1.10**	**1.14**	**1.05**

Four														
PIP	640	896	640	896	896	896	640	896	896	896	640	896	896	896
263ENC-MP3DEC	230.43	230.51	230.47	230.47	268.47	230.51	230.51	230.51	230.51	230.51	230.51	230.51	230.51	230.45
MWD	1216	2015	1723	2107	2245	1965	1332	2110	2345	1845	1216	1784	2349	1664
MPEG4	3854	4225	3991.78	4291.86	4681.37	3940.21	3712.95	4387.16	4689.37	3960.12	3728	4014.43	4789.37	3713
VOPD	4513	14,415	4652.18	15,073.18	5319.16	4900	4317	5105.74	5479.16	4773	4216	4586.54	5509.16	4237
DVOPD	12,076	11,257	16,548	17,389	19,789	11,668	9812	19,186	20,789	10,004	9638	15,239	21,089	9800
G17	47,305.92	54,251.15	51,291.91	57,166.16	59,137.92	52,507.96	43,128.41	56,912.16	59,736.92	45,622.87	39,458	53,278.83	59,536.92	40,565
G18	8513.02	9912.07	8217.17	8753.18	8924.74	8459.07	7875.83	8614.91	8827.74	6847.07	6374.26	7623.71	8767.74	6522.23
G19	8943.63	9985.12	8998.84	9318.17	9617.56	8984.38	8317.97	9418.8	9657.56	7266.12	6534.52	8235.73	9597.56	6745.89
G20	16,875.65	15,345	15,864	20,169	24,173	18,891.95	18,967.85	19,813	24,975	17,726.21	11,813.83	17,897.7	23,965	12,573.3
G21	17,854	13,172.01	15,327.16	17,569.18	20,897.19	15,931.9	10,768.18	16,891.13	22,867.19	15,701.13	106,587	148,723.78	238,573.19	121,292.33
G22	52,871.97	55,129.03	50,169.63	62,199.67	65,289.18	58,671.28	48,191.73	61,253.85	63,284.18	51,989.62	42,458.54	53,746.82	62,254.18	49,280.46
G25	20,357.69	173,242.6	147,844.18	226,108.17	296,781.71	152,145.6	138,727.82	176,538.18	197,741.71	149,812.1	104,325.23	156,873.14	207,643.71	121,129.99
G26	27,431.8	30,124.17	20,010.7	27,907.87	30,876.61	22,972.12	17,768.19	20,187.91	23,836.61	18,321.26	13,456.57	19,679.38	33,821.61	15,468.77
G27	79,732.18	77,451.6	55,781.87	62,197.19	65,781.02	57,361.02	50,189.82	56,278.76	58,771.02	53,241.18	48,957.84	54,787.34	63,861.02	49,380.91
G28	49,678.96	56,715.31	43,784.67	55,189.21	58,163.19	47,174.12	40,186.21	45,089.18	51,169.19	42,981.12	37,345.75	40,453.71	45,269.19	41,014.01
G29	491,247.81	423,506.18	323,472.98	431,872.81	456,193.17	358,215.72	281,927.76	321,926.18	353,145.17	293,158.12	248,746.53	305,647.85	352,543.17	255,441.25
Rank	1.19	1.51	1.20	1.59	1.59	1.27	1.09	1.35	1.46	1.15	**1**	1.29	1.63	1.10

Bold values indicate the entire results presented in this table

Table 4.3 Communication cost before and after re-mapping with of 25% highly utilized TSVs in 3D-NoC

Layers	Two			Four		
Benchmarks	Fully	25% TSVs	Re-map with 25% TSVs	Fully	25% TSVs	Re-map with 25% TSVs
PIP	640	1068	768	640	1996	896
263ENC-MP3DEC	230.41	350.42	230.42	230.51	830.45	230.45
MWD	1216	1648	1248	1216	2664	1664
MPEG4	3600	4732	3632	3728	4819	3713
VOPD	4119	5219	4119	4216	6339	4237
DVOPD	9544	10,692	9592	9638	10,900	9800
G17	36,295.6	49,298.9	40,198.9	39,458	46,665	40,565
G18	6094.11	6780.91	6280.91	6374.26	9622.23	6522.23
G19	6415.87	6927.12	6717.12	6534.52	9845.89	6745.89
G20	102,056	110,931	110,831	11,813.83	27,673.3	12,573.3
G21	101,256	138,380	107,280	106,587	291,392.33	121,292.33
G22	41,198.8	49,991.4	48,271.4	42,458.54	49,380.46	49,280.46
G25	101,235.32	212,415.11	112,315.11	104,325.23	191,229.99	121,129.99
G26	13,189.28	18,885.12	14,725.12	13,456.57	19,768.77	15,468.77
G27	47,129.49	69,338.17	48,238.17	48,957.84	48,470.91	49,380.91
G28	352,189.12	651,212.11	381,112.11	37,345.75	48,114.01	41,014.01
G29	349,856.14	382,412.91	242,312.91	248,746.53	259,541.25	255,441.25
Rank	**1**	**1.38**	**1.05**	**1**	**1.76**	**1.10**

Bold values indicate the entire results presented in this table

keep only 25% highly utilized TSVs. Communication cost values are recomputed such that interlayer message flow uses elevator-first algorithm (Dubois et al. 2013). Naturally, communication cost degrades. Next, it performs a remapping of tasks with the current TSV distribution. Compared to the fully connected configuration, the 25% TSV case without remapping increases communication cost by 38% and 76%, on an average for two and four layers, respectively. But, it reduces to 5% and 10% for two and four layers after remapping.

2.3   Dynamic Performance of Different Mapping and TSV Configurations

To measure the proficiency of individual mapping techniques together with different TSV configurations in the partially connected 3D-mesh-NoC-based systems, simulation has been carried out for each of the NoC-based systems using Noxim-3D simulator. Synthetic self-similar traffic has been generated, by obeying the communication requirement of tasks in the application. Tables 4.4 and 4.5 show the results of throughput, latency and average packet energy (μJ) for the benchmarks

Table 4.4 Comparison of throughput, latency and energy (μJ) of different mapping and TSV placement strategies TSV distribution for two layers

Proposed KL_Map_and_TSV_place with (Two layers)

Benchmark	Parameters	100% Fully	25% Symm.	25% Rand.	25% Intl.
263ENC-MP3DEC	Throughput	0.80	0.80	0.80	0.80
	Latency	77,461	77,461	77,461	77,461
	Pkt. energy	9.73	9.73	9.73	9.73
MWD	Throughput	0.75	0.72	0.69	0.74
	Latency	99,771	99,867	99,987	98,521
	Pkt. energy	11.68	11.89	12.05	11.76
MPEG4	Throughput	0.61	0.55	0.48	0.59
	Latency	98,059	98,969	99,567	97,531
	Pkt. energy	12.87	12.95	13.39	12.88
VOPD	Throughput	0.71	0.67	0.62	0.70
	Latency	96,523	97,831	98,679	95,321
	Pkt. energy	12.51	12.82	12.98	12.65
DVOPD	Throughput	0.65	0.62	0.59	0.64
	Latency	96,893	97,359	98,762	97,526
	Pkt. energy	11.68	11.83	11.98	11.70
G17	Throughput	0.62	0.68	0.53	0.61
	Latency	98,739	98,879	99,785	97,697
	Pkt. energy	13.69	13.88	13.99	13.71
G18	Throughput	0.69	0.64	0.60	0.65
	Latency	99,862	99,882	99,956	98,738
	Pkt. energy	11.25	11.55	11.89	11.30
G19	Throughput	0.68	0.64	0.61	0.65
	Latency	98,996	98,999	99,789	97,693
	Pkt. energy	12.68	12.75	12.98	12.72
G20	Throughput	0.70	0.66	0.63	0.69
	Latency	97,892	98,867	99,369	96,521
	Pkt. energy	13.25	13.68	13.95	13.31
G21	Throughput	0.78	0.74	0.72	0.77
	Latency	99,835	99,979	99,912	98,356
	Pkt. energy	13.01	13.35	13.85	13.21
G22	Throughput	0.61	0.57	0.51	0.59
	Latency	98,784	98,813	99,659	97,626
	Pkt. energy	13.25	13.59	13.68	13.35
G25	Throughput	0.76	0.72	0.68	0.75
	Latency	96,392	97,399	98,765	95,625
	Pkt. energy	10.59	10.72	10.85	10.62
G26	Throughput	0.75	0.71	0.68	0.74
	Latency	97,359	98,639	99,561	96,025
	Pkt. energy	11.29	11.56	11.98	11.35
G27	Throughput	0.62	0.59	0.55	0.60
	Latency	98,536	98,875	99,578	97,653
	Pkt. energy	13.56	13.89	13.98	13.62
G28	Throughput	0.65	0.61	0.57	0.64
	Latency	88,925	89,735	90,531	87,391
	Pkt. energy	12.25	12.56	12.97	12.35
G29	Throughput	0.73	0.69	0.66	0.72
	Latency	96,231	97,315	998,956	96,103
	Pkt. energy	13.68	13.95	13.99	13.70

Table 4.5 Comparison of throughput, latency and energy (μJ) of different mapping and TSV placement strategies TSV distribution for four layers

Proposed KL_Map_and_TSV_place with

Four layers	Parameters	Benchmarks: 263ENC-MP3DEC 100%	25%			Benchmarks: MWD 100%	25%			Benchmarks: MPEG4 100%	25%			Benchmarks: VOPD 100%	25%		
		Fully	Symm.	Rand.	Intl.	Fully	Symm.	Rand.	Intl.	Fully	Symm.	Rand.	Intl.	Fully	Symm.	Rand.	Intl.
	Throughput	0.78	0.78	0.78	0.78	0.72	0.66	0.64	0.72	0.58	0.53	0.50	0.56	0.70	0.66	0.64	0.69
	Latency	78,641	78,641	78,641	78,641	99,831	99,951	99,989	98,756	98,231	98,982	99,235	97,326	97,321	98,971	99,869	96,235
	Pkt. energy	9.73	9.73	9.73	9.73	10.25	10.59	10.86	10.25	12.56	12.89	13.12	12.68	11.26	11.58	11.88	11.37

	Parameters	Benchmarks: DVOPD 100%	25%			Benchmarks: G17 100%	25%			Benchmarks: G18 100%	25%			Benchmarks: G19 100%	25%		
		Fully	Symm.	Rand.	Intl.	Fully	Symm.	Rand.	Intl.	Fully	Symm.	Rand.	Intl.	Fully	Symm.	Rand.	Intl.
	Throughput	0.60	0.56	0.53	0.59	0.60	0.55	0.52	0.58	0.60	0.57	0.52	0.59	0.65	0.61	0.58	0.64
	Latency	98,535	98,882	99,785	97,321	98,529	99,621	99,923	97,569	98,231	98,895	99,876	97,325	98,756	98,856	99,789	97,325
	Pkt. energy	10.02	10.72	10.85	10.15	12.01	12.59	12.89	12.16	13.25	13.68	13.89	13.45	110.35	10.68	10.86	10.54

	Parameters	Benchmarks: G20 100%	25%			Benchmarks: G21 100%	25%			Benchmarks: G22 100%	25%			Benchmarks: G25 100%	25%		
		Fully	Symm.	Rand.	Intl.	Fully	Symm.	Rand.	Intl.	Fully	Symm.	Rand.	Intl.	Fully	Symm.	Rand.	Intl.
	Throughput	0.68	0.63	0.60	0.66	0.75	0.70	0.68	0.72	0.60	0.56	0.54	0.59	0.75	0.70	0.68	0.73
	Latency	97,325	98,125	99,625	96,239	19,235	99,623	101,239	98,321	98,321	98,976	99,215	97,521	98,312	8991	99,731	97,351
	Pkt. energy	12.56	12.98	13.00	12.65	12.11	12.52	12.96	12.35	12.01	12.87	12.99	12.31	12.01	12.68	12.98	12.38

	Parameters	Benchmarks: G26 100%	25%			Benchmarks: G27 100%	25%			Benchmarks: G28 100%	25%			Benchmarks: G29 100%	25%		
		Fully	Symm.	Rand.	Intl.	Fully	Symm.	Rand.	Intl.	Fully	Symm.	Rand.	Intl.	Fully	Symm.	Rand.	Intl.
	Throughput	0.70	0.65	0.62	0.68	0.64	0.60	0.57	0.63	0.65	0.60	0.57	0.63	0.68	0.62	0.59	0.66
	Latency	97,329	98,125	99,731	96,215	98,335	98,915	99,625	97,235	98,631	99,231	99,923	97,315	98,123	98,920	99,620	97,326
	Pkt. energy	10.25	10.62	10.98	10.27	12.59	12.82	12.96	12.65	12.21	12.69	12.98	12.51	12.43	12.79	12.92	12.51

mapped onto two- and four-layers 3D NoC. It can be noted from the table fully connected 3D NoC gives the best performance in terms of all the three factors. Using 25% TSVs results in degradation of all the three parameters. However, intelligent TSV placement with task remapping improves the solution quality compared to *25% Symm.* and the *25% Rand.* with throughput, latency and packet energy.

3 Conclusion

In this chapter, we have presented a constructive strategy to reduce the communication cost and enhance the performance of 3D NoC-based system. From the simulation results it can be noted that constructive heuristic produces better solutions for most of the application benchmarks than NMAP, PSMAP, Squeezing and KL-based heuristic. However, constructive methods work with a predetermined notion about the avenue to achieve a good solution. Almost all such methodologies work with an ordering of tasks/edges of the task graph, sorted in descending order of communication requirement. This may not work always. This has motivated us to develop an evolutionary mapping technique. The next chapter proposes a Discrete Particle Swarm Optimization (DPSO) based approach for the integrated application mapping and TSV placement problem in 3D NoC-based systems.

References

Dubois, F., Sheibanyrad, A., Petrot, F., & Bahmani, M. (2013). Elevator-first: A deadlock-free distributed routing algorithm for vertically partially connected 3D-NoCs. *IEEE Transactions on Computers, 62*(3), 609–615.

Liu, C., Zhang, L., Han, Y., & Li, X. (2011a). Vertical interconnects squeezing in symmetric 3D mesh Network-on-Chip. In *Proceeding of Asia and South Pacific Design Automation Conference (ASP-DAC)* (pp. 357–362). Piscataway: IEEE.

Murali, S., & De Micheli, G. (2004a). Bandwidth constrained mapping of cores onto NoC architectures. In *Proceeding of Design, Automation and Test in Europe (DATE)* (pp. 896–901). Piscataway: IEEE.

Sahu, P. K., Venkatesh, P., Gollapalli, S., & Chattopadhyay, S. (2011). Application mapping onto mesh structured network-on-chip using particle swarm optimization. In *Proceeding of Computer Society Annual Symposium on VLSI (ISVLSI)* (pp. 335–336). Piscataway: IEEE.

Chapter 5
A Discrete Particle Swarm Optimization Technique for Designing a 3D NoC-Based Multi-Core Systems

The evolutionary approaches based on Genetic Algorithm (GA), Ant Colony Optimization (ACO) and Particle Swarm Optimization (PSO) often perform a better exploration of the search space, compared to other heuristics, as multiple solutions do evolve simultaneously with mutual interactions between them. These exploratory strategies also need to be guided. A general observation about PSO is that it converges faster than similar techniques like Genetic Algorithms, and can work with relatively small population size. This chapter develops a PSO-based approach to solve the mapping and TSV placement problems, together. Furthermore, to check the quality of solutions generated by the proposed PSO-based approach, an exact method has been developed built around the Integer Linear Programming (ILP).

The salient features of the approach are as follows:

(a) An exact formulation to the integrated task mapping and TSV placement problem based on Integer Linear Programming (ILP) has been proposed for NoCs with two vertical layers.

(b) A Discrete Particle Swarm Optimization (DPSO) based approach has been developed for integrated application mapping and TSV placement problem to minimize the overall communication cost.

(c) The basic PSO formulation has been augmented by (1) an efficient random number generator, (2) inversion mutation (IM) operation, (3) deterministically generating a part of the initial population for PSO and (4) running multiple PSO.

(d) For any stage of PSO, the initial population generation is not fully random. A good number of particles have been created using fast search techniques built around the constructive mapping algorithm proposed in our earlier works. This has enabled our PSO to explore the promising regions of search space better.

(e) We have used a multi-stage PSO. The local and global best information of ith stage are passed to the particles in $(i + 1)$th stage. This ensures faster convergence and improved quality of solution for the successive stages.

© Springer Nature Switzerland AG 2020
K. Manna, J. Mathew, *Design and Test Strategies for 2D/3D Integration for NoC-based Multicore Architectures*, https://doi.org/10.1007/978-3-030-31310-4_5

1   ILP Formulation for TSV Placement and Application Mapping

An ILP formulation for the TSV placement and mapping problem has been described here. The parameters and variables used in the ILP formulation are noted in Table 5.1.

1.1   *Objective Function*

The objective of the work is to minimize the communication cost by placing the TSVs at suitable positions of the 3D NoC-based system and performing a mapping of tasks to cores. The objective function can be written as

$$Minimize: \sum_{\forall l \in E} Bw_l \times \left(\sum_{\forall (\mu_s, \mu_t) \in U,\ k \in \{valid\ TSV\}} D^{T_k}_{\mu_s \mu_t} \times Y_{lk} \right) \times P^{\mu_s \mu_t}_l \quad (5.1)$$

where k represents the index of a core and the vertical connection (TSV) of kth router is denoted as T_k. This work considers two layers (i.e., $l = 2$) and all vertical connections to be bidirectional in nature. The value of k can range from 1 to the number of routers present per layer. For example, if we have 12 cores and the number of routers per layer is 6 (the cores being equally distributed in different layers), k can take any value between 1 and 6.

The distance between routers, present in different layers, can change depending upon the choice of TSV used. Hence, for every pair of cores/routers, all possible

Table 5.1 Variables used for ILP formulation

Variables	Definitions
$m^{\mu_s}_{c_i}$	= 1, if ith task c_i is mapped to sth core μ_s
	= 0, otherwise
$P^{\mu_s \mu_t}_l$	= 1, if communication path exists between cores μ_s & μ_t for a mapped edge l of the task graph
	= 0, otherwise
T_k	= 1, if kth router has a vertical link
	= 0, otherwise
$D^{T_k}_{\mu_s \mu_t}$	Pre-computed distance between cores μ_s and μ_t, given kth router has TSV, that is, $T_k = 1$
Y_{lk}	=1, if $T_k = 1$ and kth router comes in the path for a mapped edge l. This depends on the routing algorithm followed.
	=0, otherwise
Bw_l	= Bandwidth requirement of the edge l of the task graph.

distances have been precomputed in $D_{\mu_s \mu_t}^{T_k}$ and provided in the ILP formulation. The appropriate distance will get selected based on the value of the variable T_k. Suppose, for an edge $l = (c_i, c_j)$ in the task graph, task c_i is mapped to core u_s, and c_j to u_t, located in different layers. The routing path between cores u_s (source) and u_t (destination) has been decided by the routing algorithm. Now, such path contains a router having vertical connection (i.e., $T_k = 1$). The usage of such TSV is captured by a binary variable Y_{lk}. The value $Y_{lk} = 1$ signifies that kth router has a vertical connection and it comes in the routing path for the mapped edge l, and 0 otherwise.

1.2 Constraints

The following set of constraints have been framed to solve the TSV placement and mapping problem in 3D NoC-based systems.

(a) **TSV usage constraints**

$$Y_{lk} \leq T_k \tag{5.2}$$

for all edge l of task graph and all k of TSV positions

$$\sum_{k=1}^{\substack{No.\ of\ routers\ per\ layer}} Y_{lk} = 1 \tag{5.3}$$

The first constraint ensures that the path distance with a chosen TSV k is considered for a task graph edge l only when that TSV is present ($T_k = 1$). The second constraint ensures that an edge l is mapped considering only one TSV at most. For example, suppose, we have TSVs at both first and third locations (i.e., $T_1 = T_3 = 1$). While mapping an edge l, its distance will include either T_1 or T_3. However, if an edge l in the task graph is mapped to the cores in the same layer, neither T_1 nor T_3 will come in the path between the cores. In this case, any of Y_{l1} or Y_{l3} can become 1.

(b) **Mapping constraints**

$$\forall \mu_s \in U, \sum_{c_i \in C} m_{c_i}^{\mu_s} \leq 1 \tag{5.4}$$

$$\forall c_i \in C, \sum_{\mu_s \in U} m_{c_i}^{\mu_s} = 1 \tag{5.5}$$

The first inequality implies that any router has at most one core attached to it. Constraint (5.5) ensures that each task has to be mapped onto only one core.

These constraints guarantee that no two tasks are mapped to one core and each core gets attached to a single router.

(c) **Constraints for task graph edges**

$$\forall l \in E, \forall (\mu_s, \mu_t) \in U \left[P_l^{\mu_s \mu_t} \geq m_{c_i}^{\mu_s} + m_{c_j}^{\mu_t} - 1 \right] \tag{5.6}$$

where c_i is source of lth edge and it is mapped to core μ_s and c_j is destination of lth edge and is mapped to core μ_t

$$\forall l \in E, \forall (\mu_s, \mu_t) \in U \left[P_l^{\mu_s \mu_t} \leq (m_{c_i}^{\mu_s} + m_{c_j}^{\mu_t}) \div 2 \right] \tag{5.7}$$

where c_i is source of lth edge and it is mapped to core μ_s and c_j is destination of lth edge and its mapped to core μ_t.

Each edge present in the task graph would be mapped onto a path in the topology graph. It can be ensured by the inequalities (5.6) and (5.7). The variable $P_l^{\mu_s \mu_t}$ will be 1, when edge l of the task graph is mapped to a path between core μ_s and μ_t.

(d) **Technological constraints**

$$\sum T_k = n/4 \tag{5.8}$$

where n is the number of routers per layer

$$T_k + T_{k+1} = 1 \tag{5.9}$$

for k not being a router in last column

$$T_k + T_{k+n} = 1 \tag{5.10}$$

for k not being a router in last row, n being the number of routers per layer.

Due to the technological constraints, we assumed that the number of vertical links (TSVs) in each layer can be less than or equal to 25% of the total routers present per layer. This criteria can be ensured by the constraint (5.8). The constraints (5.9) and (5.10) ensure that TSVs are not placed within one hop distance of each other.

The tool CPLEX (Cplex 2013) has been used to solve the formulated ILP and get optimum solution. However, except for very small NoCs, it takes a huge amount of CPU time to arrive at the solution. Hence, in the following, a Particle Swarm Optimization (PSO) based technique and its variants have been proposed to find the solutions for bigger NoCs, producing results within a reasonable amount of CPU time.

2 PSO Formulation for TSV Placement and Application Mapping

Particle Swarm Optimization (PSO) (Kennedy and Eberhart 1995) is a population-based stochastic optimization technique developed by Eberhart and Kennedy, encouraged by the social behaviour of bird flocking and fish schooling. In this technique, multiple solutions are present at any instant of the optimization phase. These solutions help each other to evolve themselves by sharing their experiences to achieve close to an optimum solution. Each of these solutions is called a *particle*. A particle moves through the solution space according to its own experience as well as the experience of the fellow particles. The proficiency of a particle is measured by its fitness. PSO has been applied successfully to solve many optimization problems in both continuous and discrete domains (Wang et al. 2003). This has motivated us to look for a discrete PSO (DPSO) formulation of the integrated TSV placement and task mapping problem for the 3D-mesh-based NoC with limited vertical connections. Apart from developing a PSO, we have augmented it in several ways as discussed in Sect. 2.2.

The position of a particle, in an n-dimensional search space at the kth iteration can be represented as $p_k = < p_{k,1}, p_{k,2}, \ldots, p_{k,n} >$. Let, p_k^i denote the position for the ith particle. For the ith particle, let its local best position be represented by $pbest^i$, corresponding to the best fit position that the particle has seen so far over the generations. Similarly, the global best particle of kth generation may be represented by $gbest_k$. The particles can *fly* over the solution space through generations. The new position can be calculated as

$$p_{k+1}^i = \left(k_1 * I \oplus k_2 * \left(p_k \rightarrow pbest^i \right) \oplus k_3 * (p_k \rightarrow gbest_k) \right) p_k^i \qquad (5.11)$$

In this equation, $m \rightarrow n$ denotes the minimum length sequence of swapping to be applied on m to transform it to n. For example, let, $m = < 1, 3, 4, 2 >$ and $n = < 2, 1, 3, 4 >$, $m \rightarrow n = < swap(1, 4), swap(2, 4), swap(3, 4) >$. The operator $\oplus$ denotes the fusion operator. Two swap sequences are applied one after another, i.e., m followed by n, for the operation $m \oplus n$. Each particle evolves over the generations based on inertia, self-confidence and swarm confidence. In expression (5.11), the corresponding factors are denoted by the constants k_1, k_2 and k_3, respectively. The quantity $k_i * (m \rightarrow n)$ means that the swaps in the sequence $(m \rightarrow n)$ are to be applied with probability k_i. The identity swap sequence is represented by $I = < swap(1, 1), swap(2, 2), \ldots, swap(n, n) >$. It corresponds to the inertia of the particle to maintain its initial configuration. The final sequence of swaps equivalent to $(k_1 * I \oplus k_2 * (p_k \rightarrow pbest^i) \oplus k_3 * (p_k \rightarrow gbest_k))$ are to be applied on particle p_k^i to create p_{k+1}^i.

The convergence criteria for the DPSO is given by Guilan et al. (2008)

$$\left(1 - \sqrt{k_1} \right)^2 \leq k_1 + k_3 \leq \left(1 + \sqrt{k_1} \right)^2 \qquad (5.12)$$

Accordingly, we have tried with various values of k_1, k_2 and k_3. The results reported in this work consider the values of $k_1 = 1, k_2 = 0.04$ and $k_3 = 0.02$. Next, we present the particle structure used by us in the integrated TSV position selection and core mapping problem.

2.1   Particle Formulation and Fitness Function

2.1.1   Particle Structure

To solve the integrated problem of task mapping and TSV position selection using PSO, the particle structure has been made to have two parts in it. The first part corresponds to the mapping, while the second part indicates the routers having a vertical connection via TSV. For this, the routers have been numbered in increasing order from the lowest to the highest layer. Within a layer, the router numbers are assigned in a row-wise manner, starting from the top-left upto the bottom-right corner. Figure 5.1a shows the numbering scheme followed for a $3 \times 3 \times 2$, 2-tier NoC having a total of 18 routers. The *task mapping part* of the particle is a permutation of task numbers, identifying the task mapped to a core. For example, Fig. 5.1b shows the particle structure corresponding to the NoC mapping in Fig. 5.1a. Task16 gets mapped to core1, task 13 to core 2, and so on.

For the TSV part, it has been assumed that the routers at similar positions are of the same type in each layer. That is, if router r in layer 1 is a 3D router with a TSV connection to corresponding router r' in layer 2, $\dot r$ is also having a TSV connection to router r'' in layer 3, and so on. Such an assumption is justified as

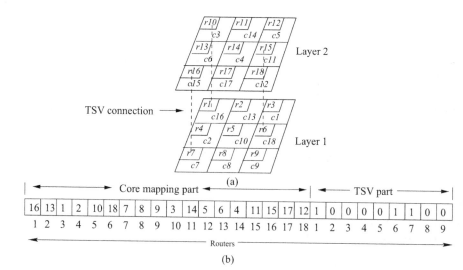

Fig. 5.1 Particle structure for application mapping and TSVs placement problem. (**a**) A vertically-partially-connected 3D-mesh-NoC. (**b**) Particle structure for the configuration in (**a**)

TSV geometry will not allow two neighbouring routers in any layer to have TSV connections. Thus, it is better to put 3D routers at the same positions in each layer. The TSV part of the particle has been formulated as a bit-array of size equal to the number of routers at layer l (or any other layer). A bit being '1' indicates that the corresponding routers in each layer are 3D in nature with TSV connections. A '0' indicates the router to be 2D without TSV connection. The TSV-placement constraints, such as $x\%$ routers in each layer can have TSV and that the minimum d-hop distance must be maintained between neighbouring routers having TSV, are not incorporated into the particle. However, the constraints have been considered while determining the 3D NoC structure corresponding to the particle. For example, assume that the first bit of TSV part of a particle is '1'. This router will have a TSV. During placement of TSVs for other routers (i.e., routers having '1' bit in the TSV part of the particle), it will consider the constraint. Such a router can have TSV, only if it satisfies both the constraints. Figure 5.1b shows a full particle structure. For the purpose of implementation, a particle has been considered to be a single array with appropriate care taken regarding the values contained in its cells. The fitness of each particle can be evaluated using expression (5.13)

$$CommCost = \sum_i \sum_j \left(BW\left(c_i, c_j\right) \times Dist\left(r_i, r_j\right) \right) \qquad (5.13)$$

where $BW(c_i, c_j)$ denotes the bandwidth requirement between tasks c_i and c_j. $Dist(r_i, r_j)$ is the hop count in a shortest path between cores r_i and r_j, to which the tasks c_i and c_j have been mapped.

2.1.2 Local and Global Best Particle

Each particle has an associated local best (*pbest*) which is a configuration with the minimum communication cost (defined by Eq. (5.13)), among all configurations that the particle has seen so far, in the evolution process. On the other hand, global best (*gbest*) is a particle having the best (minimum) communication cost for a generation which has been calculated from the set of local bests. The local as well as the global best particles control the evolution of each particle. The local and global best particles are updated if the corresponding fitness values in the current generation are less than the values until the previous generation.

2.1.3 Evolution of the Generation

Particles evolve over generations to create new particles which are expected to produce better solutions. The initial population is created randomly and the fitness values of individual particles are determined. The local best (*pbest*) of each particle is initialized to itself. For a new generation, particles are created through a series of *swap* operations, explained next.

2.1.4   Swap Operator

Swap operation takes two indices, say i and j, of the particle p as input and creates a new particle p_1. The particles p and p_1 are same excepting that the positions i and j of p are exchanged in p_1. Care has been taken to disallow swapping between task part and TSV part of a particle.

Let p be a particle as shown in Fig. 5.2a. The swap operator $SO(3,5)$ exchanges the values at positions 3 and 5 in p to generate a new particle as shown in Fig. 5.2b.

2.1.5   Swap Sequence

It is a sequence of swap operators. For example, a swap sequence $SS =< \{SO(7, 1), SO(4, 3)\} >$ creates particle, P_{new}, by applying the operations on particle P in two steps. Figure 5.3a represents the particle P. Applying $SO(7,1)$ on P creates an intermediate particle, P_{mod}, shown in Fig. 5.3b. The swap $SO(3,4)$ on particle P_{mod} results in the new particle noted in Fig. 5.3c.

For the evolution of a particle, first the swap sequences are identified to align it to its local best and the global best. The sequences are applied with some probabilities corresponding to the confidence factors. For our formulation, we have used the confidence factors to be 0.04 and 0.02, respectively, for local and global best alignment.

Fig. 5.2 An example of swap operation. (**a**) A particle before applying the *swap* operation. (**b**) The particle after applying the *swap* operation

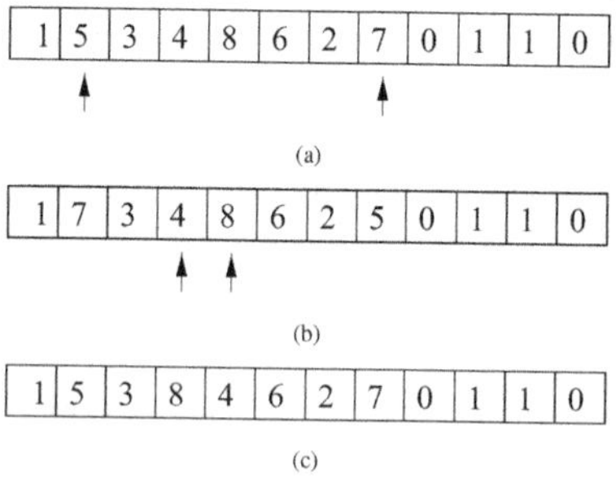

Fig. 5.3 An example of swap sequence operation. (**a**) A particle before applying the *swap* operation. (**b**) The particle after applying a *swap* operation. (**c**) The particle after applying the next *swap* operation

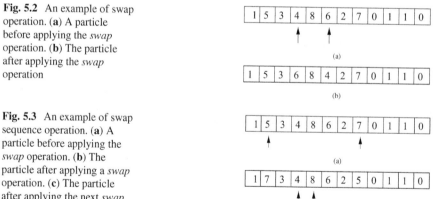

2.2 Augmentation to the Basic PSO

To achieve a better solution from the PSO technique presented so far, the following augmentations have been incorporated.

2.2.1 Usage of Better Random Number Generator

PSO algorithm depends heavily on the quality of random number generator available in the systems. The available C-library routines for random number generation, `rand()` and `rand48()` use Linear Congruential Generator (LCG) (Saito and Matsumoto 2008). We propose to use the thread-safe single instruction multiple data-oriented fast Mersenne twister (SIMT) pseudorandom number generation technique (Saito and Matsumoto 2008). The technique provides several advantages over LCG. In particular, it has larger period (upto $2^{216091} - 1$), compared to $(2^{31} - 1)$ for LCG (Saito and Matsumoto 2008), better equidistribution and quick recovery from 0-excess initial state. Compared to other statistically reasonable generators, it is faster and useful, when huge random values are required (Tian and Benkrid 2009). The SIMT has passed several statistical testing including the diehard test of Marsaglia and the load test of Hellekalek and Wegenkittl (Matsumoto and Nishimura 1998). The effect of using SIMT, instead of LCG has been demonstrated in Sect. 3.3.

2.2.2 Inversion Mutation (IM)

The PSO generally performs better than Genetic Algorithm (GA) (Guilan et al. 2008). However, one important drawback of PSO, as compared to GA, is the absence of a mutation operator that can bring sudden changes into a solution, thus possibly exploring a promising unexplored part of the search space. When PSO is found to be not improving over a fixed predefined number of generations, a mutation operation may take it out of a probable local optima. In this light, we have introduced an *inversion mutation* operator. To apply this on a particle, we follow the procedure noted next. First, a break-point is randomly generated for the task part of the particle. The portion from this break-point to end is inverted and joined at the end of the part before break-point. Next, the same is performed for the TSV part. Figure 5.4 shows the operation of such an inversion mutation. Its impact on solution quality has been shown in Sect. 3.3.

2.2.3 Multiple PSO

In any population-based search technique, exploration and exploitation are the two properties that can be used to control the quality of the solution. The techniques can also be used in PSO. In the exploration phase, different regions of the search space

Fig. 5.4 An example of Inversion Mutation operation. (**a**) A particle before applying the *mutation* operation. (**b**) The particle after applying the mutation operation

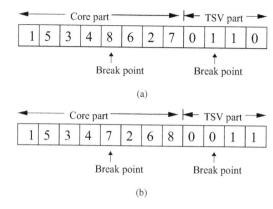

are explored, whereas, the exploitation process checks for local optima around the globally explored points. In the initial portion of a PSO run, it performs more of exploration. However, through the evolution process, the particles start converging, making more of exploitation. To balance the exploration and exploitation process in a multiple swarm based optimization techniques, several strategies can be found in the literature (Rohler and Chen 2011; Chen and Montgomery 2011). Among these strategies, locust swarm (Rohler and Chen 2011) is based upon *devor and move on* strategy. In this strategy, if a sub-swarm has found a local optima, a set of scouts are deployed to explore the new potential regions. Furthermore, the scouts are guided by intelligence accumulated by the earlier sub-swarm. In our work, we have utilized a similar such technique for better exploration of the search space, as detailed next.

In our proposed augmentation, PSO has been run several times to improve upon the global best solution. Suppose that at the end of nth run of PSO, the local best for kth particle be $pbest_n^k$, and the global best be $gbest_n$. In the $(n + 1)$th pass of PSO, it starts with a new set of particles. However, the local and global best information is transferred from the nth to the $(n + 1)$th PSO. The maximum number of the PSO runs to be executed has been restricted as follows:

- A user-defined value for the maximum number of PSO runs. In this experiment, it has been taken as 200.
- The global best fitness does not change in the last 20 PSO runs.

2.2.4 Initial Population Generation

For a task graph with n tasks mapped onto a 3D-mesh having n routers distributed over m layers, the total number of possible mappings and TSV positions can be $n!$ and $(2^{n/m})$, respectively. Exploration of the potential region of this enormous search space depends to a great extent on the initial set of particles. We have used the deterministic initial population generation technique, described in Chap. 4, to help in the process. The strategy proposed in Chap. 4 can generate a number of solutions equal to the number of core in the NoC, quite fast. However, it works only for a fully connected 3D mesh. To restrict the number of TSVs, we have generated a number of 3D mesh architectures with randomly placed TSVs. The total number of TSVs has been restricted to 25% of total available positions. Also, TSVs are placed at least two hops away from each other. Next, the algorithm suggested in Chap. 4 has been utilized to get a number of mapping solutions. For each core in the NoC, a solution is generated by starting the mapping process at that core. The best one among them, along with the associated TSV positions contributes to the creation of one intelligent particle to be included in the initial population. The process has been repeated to create a certain number of intelligent particles. The rest of the particles are generated randomly.

2.3 PSO-Based Application Mapping and TSV Placement Algorithm

The entire PSO engine has been described in Algorithm 5.1. In this algorithm, each *Particle* represents a configuration of mapping of tasks into the core of NoC topology. Such configuration represents a position in the solution space. The algorithm generates few intelligent and remaining random particles, a total of *N Part* number of particles in the initial generation of PSO (line 10). Few intelligent configurations are generated by using the strategy, described in Chap. 4. The local best of individual particles is set to the particle itself. The global best particle is set to be the best configuration found in the local best configuration set (line 20). Each configuration can be judged based on the expression (5.13). Here, *NPart* is the number of particles, while *MGEN* is the maximum number of generations. The number of PSO runs is represented by *MPSO*.

After generating the initial configurations and finding the global best configuration at first generation, the particles are evolved using *UpdatePart()*. Each particle evolves by sharing the experience of its local as well as global best of the generation with a random probability. The *swap* sequences are generated based on the particle, local and global best configurations. New particle or configuration is generated by applying those swap sequences onto the particle with a certain probability. Furthermore, the local best of each particle gets changed by comparing the fitness between its current *lbest* and the new particle (line 17). After each generation, the PSO engine checks the *BestFitness* value. It changes the current generation to the initial generation if it gets better fitness than the earlier one. Otherwise, it keeps count of the generations. The *inversion mutation* (IM) (line 29) is applied onto each particle, if generation count reaches up to (*MGEN*) and *IM* is *true*. The PSO engine creates new particles by assigning random configurations using *Random()*, whereas local best for those particles are passed from the earlier PSO run using *CopyLbest()*, when the generation count reaches to its maximum value (lines 32–35). The PSO engine stops when multiple PSO count reaches the maximum value.

Algorithm 5.1 PSO-based application mapping and TSV placement algorithm

---

Input: Task graph C, Topology graph T,
Output: Mapping of C to T along with position of TSVs in T satisfying restrictions on TSV
  numbers and places.

1: Set $MGEN$, $MPSO$ and $NPart$
2: Set TSVs constraints
3: $BestFitness \leftarrow \infty$
4: **for** m from 0 to $MPSO$ **do**
5:     $IM \leftarrow true$
6:     $BeforeIMBestFitness \leftarrow \infty$
7:     **while** $IM = true$ **do**
8:         $gen \leftarrow 0$
9:         **while** $gen < MGEN$ and $IM = true$ **do**
10:             **for** p from 0 to $NPart$ **do**
11:                 **if** $gen = 0$ and $m = 0$ **then** //Initialization
12:                     $Particle_p \leftarrow \{CommIntelligent(), Random()\}$
13:                     $Particle_p^{pbest} \leftarrow Particle_p$
14:                 **else**//Update particle
15:                     UpdatePart($Particle_p$, $Particle_p^{pbest}$, $Particle^{gbest}$, $ProbRandom_{pbest}$, $ProbRandom_{gbest}$)
16:                     Compute fitness of $Particle_p$ using expression (5.13)
17:                     UpdateLBest($Particle_p^{pbest}$, $Particle_p$)
18:                 **end if**
19:             **end for**
20:             Compute global best particle $Particle^{gbest}$ from the set of particles $Particle^{pbest}$
21:             Update the $BestFitness$ as global best as per the requirement and reset the gen counter
22:             **if** $BeforeIMBestFitness = BestFitness$ **then**

23:                 $IM \leftarrow false$
24:             **end if**
25:             $gen \leftarrow gen + 1$
26:         **end while**
27:         **if** $IM = true$ **then**
28:             $BeforeIMBestFitness \leftarrow BestFitness$
29:             Perform Inverse Mutation (IM) at random position of each particle of the last generation
30:         **end if**
31:     **end while**
32:     **for** p from 0 to $NPart$ **do**
33:         $Particle_p \leftarrow \{Random()\}$
34:         CopyLbest($Particle_p^{pbest}$)
35:     **end for**
36: **end for**

---

3 Experimental Results and Analysis

This section presents the experimental results on different NoC benchmarks and compares these with ILP, other existing techniques and the proposed technique in previous chapters (KL and constructive) for a number of NoC benchmark applications. The simulation environment is the same as that in Sect. 5.

3.1 Communication Cost Comparison Between ILP and PSO

To check the optimality of the proposed approach for TSV placement and mapping problem, we first compare the ILP with PSO results. Table 5.2 shows the TSV placement and mapping results using ILP and PSO. For applications PIP, S1 and S2, PSO could obtain the same solution, reported by ILP. For other applications, ILP could not start or complete due to the creation of a large number of constraints. The CPLEX (Cplex 2013) tool has been used to solve the formulated ILP. Both ILP and PSO have been implemented on a Dell PowerEdge T410 system with 8 cores (Intel Xeon processor, E5606@2.12 GHz), 64 GB RAM. The capacity of PSO to reach the optimal results found from ILP gives us confidence about its quality.

3.2 Impact of Initial Population Generation

To improve the solution quality, the proposed method augments the initial population generation process of the basic PSO. That is, some percentage of the total number of particles has been taken from a fast deterministic heuristic, explained earlier (Sect. 2.2.4). The impact of this augmentation on the solution quality, with partially connected 3D-mesh-NoC with two layers, has been shown in Table 5.3. The last row of the table notes the average percentage improvements achieved

Table 5.2 Comparison of communication cost between ILP and PSO

PIP	
Mapping algorithms	Comm. cost (Hops × BW)
Proposed PSO	768
ILP	768
S1 (6 cores)	
Proposed PSO	384
ILP	384
S2 (6 cores)	
Proposed PSO	1224
ILP	1224

Table 5.3 Communication cost for intelligent initial population

Benchmarks	Intelligent particles				
	0%	1%	5%	10%	20%
G17	46,412.4	37,730	37,565	37,565	37,565
G18	9620.64	6679	6193	6153	6153
G19	9215	6689	6445	6414	6414
G20	117,978.8	105,584	105,468	105,468	105,468
G25	160,396.69	104,719.25	103,819.67	103,819.67	103,819.67
G26	28,517.03	15,217.35	13,279.89	13,269.79	13,269.79
G27	74,391.23	48,749.25	48,464.99	48,460.58	48,460.58
Avg. Imp.	–	29%	31.27%	31.38%	31.38%

Table 5.4 Communication cost comparison of different augmentation techniques

Benchmarks	Basic PSO	Basic PSO w-IM only	Basic PSO w-SIMT only
G28	499,896.19	493,796.14	486,559.15
G29	398,451	397,123	386,714

via incorporation of different percentages of intelligent particles in the initial population, over a completely random initial population. It shows that incorporation of 5% intelligent particles in the initial population can be a good augmentation. The solution quality does not improve significantly as the percentage of intelligent particles is increased from 5 to 20%. The effect of augmentations like IM and randomness of random number generator into the basic PSO have been presented in this section. The corresponding results have been noted in Table 5.4 for the applications G28 and G29. The second column notes the results of basic PSO without any augmentation. The third and fourth columns show the results of incorporation of inverse mutation and SIMT into this basic PSO. As it can be observed from the table, better results have been achieved using IM and SIMT. This establishes the suitability of the proposed augmentation strategies for improving the solution quality.

3.3   Effect of Inversion Mutation (IM) and Random Number Generator

3.4   Comparison with Existing Works

This section compares the experimental results of the current approach with some of the recent approaches reported in the literature. The corresponding results have been presented in Table 5.5. From the performance viewpoint, the best possible is the 3D mesh architecture with each router have a vertical connection (i.e., 100% TSVs) (Feero and Pande 2009). We have considered such a fully connected 3D-mesh-NoC as an instance and mapped each benchmark onto that architecture using our proposed augmented multi PSO (AMPSO) based technique to evaluate the performance. The corresponding results have been noted in the third column of Table 5.5, labeled as *AMPSO*. The TSV foot-print in such case is very high. *Rand.* in column 4 corresponds to a random distribution of 25% TSVs with a restriction that no two adjacent routers are having TSVs. The scheme suggested in Liu et al. (2011a) makes four adjacent routers to share one TSV located at the centre of the cluster, which has been labeled as *Squeezing* in Table 5.5. Thus, it uses 25% routers to have a vertical connection. We have extended the NMAP (Murali and Micheli 2004a) to work with 25% intelligently placed TSVs which have been marked as *Extnd.-NMAP* in Table 5.5. We have also compared the proposed technique with the works proposed in the last two chapters which are labeled as *KL* and *Constructive* in the table. The columns marked as *Single PSO* and *Multiple PSO* correspond to the cases in which PSO has been run only once, or several times, as noted in Sect. 2.2. Furthermore, under those columns *w/o augmntn* represents the situation in which augmentations suggested in Sect. 2.2 have not been incorporated. On the other hand, the columns marked as *w augmntn* contain the results in which all the augmentation have been used. Taking the *AMPSO* results as unity or reference, the proposed approach in the last column requires only 3.8% more communication cost, on an average, compared to 44.2%, 41.3%, 23.4% 17.9% and 9.3% more for randomly placed, (Liu et al. 2011a; Murali and Micheli 2004a), KL and Constructive, respectively, for partially connected 3D-mesh based NoC having two layers. For four layers, it takes 9.9% more communication cost, on an average, compared to 85.2%, 54.4%, 38.6%, 34.2% and 19.4% more for randomly placed, (Liu et al. 2011a; Murali and Micheli 2004a), KL and Constructive. Therefore, the proposed strategy can produce a better solution than other contemporary approaches available in the literature.

3.5 Dynamic Evaluation of Proposed Solutions

For a better understanding of the impact of TSV placement and mapping, we have next simulated each of the NoC-based systems. Synthetic self-similar traffic has been generated, guided by the communication requirement of the tasks in an application. It has been reported in Varatkar and Marculescu (2004) that on-chip modules in typical video and networking applications follow bursty traffic. The *Noxim* (Vincenzo et al. 2016) simulator has been used to simulate the NoCs. We have incorporated both the routing algorithms—elevator-first (Bahmani et al. 2012; Dubois et al. 2013) and modified-elevator-first (Lee and Choi 2013) in Noxim. The TSV positions, generated from the proposed methodology, have been provided to Noxim. To make such kind of facility into the Noxim, it has been extended accordingly. The network throughput (*Th.*), average latency (*Lat.(cycle)*), and average packet energy for the benchmarks, by running the simulation for 200,000 clock cycles, have been noted in Table 5.6. It may be observed that the

Table 5.5 Comparison of communication cost with existing works

Layers	Benchmarks	TSV Used 100% AMPSO	25% Rand.	Squeezing (Liu et al. 2011a)	Extnd.-NMAP (Murali and Micheli 2004a)	KL	Constructive	Proposed PSO and its variants Single PSO w/o augmntn	Single PSO w augmntn	Multi PSO w/o augmntn	Multi PSO w augmntn
Two	PIP	640	896	768	896	768	768	768	768	768	768
	263ENC-MP3DEC	230.41	268	230.21	230.21	230.99	230.42	230.43	230.43	230.41	230.43
	MWD	1216	1260	1248	1255	1252	1248	1248	1248	1248	1216
	MPEG4	3567	3768	3714	3672	3706	3632	3632	3632	3632	3632
	VOPD	4119	4157	4157	4279	4189	4119	4157	4141	4151	4119
	DVOPD	9528	10,384	10,307	10,404	9726	9592	10,032	9592	10,016	9554
	G17	36,295.6	46,620.8	50,626	39,387.91	38,576.63	40,198.9	46,412.4	37,565	39,974	35,375.93
	G18	6094.11	10,159.9	9814.02	7579.85	6847.48	6280.91	9620.64	6279	8184	6094.11
	G19	6102.65	9625	9930.1	8886.84	7231.05	6717.12	9215	6689	8837.5	6430.65
	G20	96,187.35	118,456	117,380.8	117,768.57	114,297.94	110,831	117,978.8	105,584	111,986	103,727.15
	G21	94,681.89	134,817	129,985	111,384.42	116,985.25	107,280	123,884.3	113,489	109,516	99,511.17
	G22	39,600	54,366.2	54,354	51,509.32	45,223.15	48,271.4	52,905.8	50,323	45,943	42,167.82
	G25	95,315.77	171,234.15	166,271.15	145,123.25	134,161.25	112,315.11	160,396.69	103,819.67	120,021.79	99,815.93
	G26	12,988.05	29,987.09	29,231.51	19,201.61	20,613.78	14,725.12	28,517.03	13,279.89	20,097.92	13,118.97
	G27	46,028.05	75,848.18	75,128.27	53,489.71	52,374.61	48,238.17	74,391.23	48,464.99	55,859.67	47,121.62
	G28	340,893.63	556,321.12	551,751.31	442,131.51	432,371.81	381,112.11	492,880.19	391,165.13	406,676.23	376,313
	G29	218,147.19	397,620.18	396,121.45	312,261.48	305,166.55	242,312.91	363,630	229,716.31	258,872.25	222,481
	Rank	**1**	**1.442**	**1.413**	**1.234**	**1.179**	**1.093**	**1.358**	**1.079**	**1.185**	**1.038**

Four	PIP	640	896	896	896	896	896	896	896	896	896
	263ENC-MP3DEC	230.43	230.51	230.47	230.51	230.51	230.45	230.45	230.45	230.45	230.45
	MWD	1216	1699	2015	1965	1845	1664	1728	1664	1664	1664
	MPEG4	3633	3912	4025	3940.21	3960.12	3713	3752	3752	3752	3713
	VOPD	4119	4310	4415	4900	4773	4237	4237	4237	4237	4237
	DVOPD	9560	9912	11,257	11,668	10,004	9800	11,085	9832	10,496	9768
	G17	32,922.35	42,189.9	54,251.15	52,507.96	45,622.87	40,565	44,305.92	36,566.99	42,224.91	36,565
	G18	6153.1	6422.15	9912.07	8459.07	6847.07	6522.23	9804.02	6222.23	8385.52	6222.23
	G19	6247.98	6912.1	9985.12	8984.38	7266.12	6745.89	9950.26	6575.89	6690.79	6545.89
	G20	96,679.95	153,118.15	123,450.19	188,913.95	177,260.21	125,737.3	19,531.92	116,290.52	111,666.7	105,737.3
	G21	93,863.07	131,213.15	131,721.01	159,312.89	157,012.13	121,292.33	128,547.79	111,292.33	121,444.17	101,035.16
	G22	37,903.21	49,172.5	55,129.03	58,671.28	51,989.62	49,280.46	55,307.56	45,259.43	51,738.26	42,280.46
	G25	89,079.84	142,140.12	173,242.61	152,142.61	149,812.12	121,129.99	148,357.69	114,208.68	122,315.89	99,126.77
	G26	12,953.47	15,311.38	30,124.17	22,972.12	18,321.26	15,468.77	27,823.64	14,520.25	20,684.07	13,402.08
	G27	42,210.77	49,231.12	77,451.6	57,361.02	53,241.18	49,380.91	79,732.18	54,027.32	57,585.57	46,380.91
	G28	32,909.72	381,215.01	56,715.31	47,174.12	42,891.12	41,014.01	44,306.13	39,324.78	40,250.19	35,014.01
	G29	210,267.34	272,510.91	423,506.18	35,821.72	293,158.12	255,441.25	336,384.47	260,493.25	275,643.72	225,441.25
	Rank	**1**	**1.852**	**1.544**	**1.386**	**1.342**	**1.194**	**1.368**	**1.161**	**1.222**	**1.099**

Bold values indicate the entire results presented in this table

Table 5.6 Throughput, latency and energy consumption of different strategies

Two number of layers

TSV used — Routing algorithm used

Benchmarks	Parameters	100%	25%		100%	25%		100%	25%		100%	25%	
		XYZ	Δ	Δ'	XYZ	Δ	Δ'	XYZ	Δ	Δ'	XYZ	Δ	Δ'
		263ENC-MP3DEC			MWD			MPEG4			VOPD		
	Th.	0.80	0.80	0.71	0.78	0.78	0.75	0.65	0.64	0.60	0.75	0.67	0.65
	Lat.(cycle)	77,461	76,480	77,843	98,259	98,260	98,304	96,342	96,260	97,014	93,456	92,340	93,550
	Avg. pkt. enrgy. (in uJ)	9.73	9.83	11.22	10.84	10.84	11.06	8.54	11.67	12.21	12.24	12.29	12.98
		DVOPD			G17			G18			G19		
	Th.	0.70	0.69	0.65	0.66	0.65	0.62	0.72	0.71	0.69	0.70	0.69	0.68
	Lat.(cycle)	94,375	93,378	94,520	97,942	97,679	98,768	98,850	97,860	98,860	98,974	98,175	98,974
	Avg. pkt. enrgy. (in uJ)	11.03	11.26	11.35	13.10	13.49	13.55	10.79	10.83	10.83	11.08	11.17	11.18
		G20			G21			G22			G25		
	Th.	0.74	0.73	0.71	0.86	0.85	0.82	0.68	0.67	0.66	0.81	0.79	0.76
	Lat.(cycle)	98,970	98,424	99,433	99,802	98,253	99,300	98,954	98,558	99,566	81,923	80,512	82,261
	Avg. pkt. enrgy. (in uJ)	12.07	12.48	13.65	12.24	12.29	12.09	12.08	12.81	13.38	10.25	10.92	11.25
		G26			G27			G28			G29		
	Th.	0.78	0.76	0.74	0.68	0.65	0.62	0.70	0.68	0.65	0.75	0.73	0.70
	Lat.(cycle)	81,235	80,111	80,971	84,231	83,192	83,912	85,427	83,129	83,997	83,125	80,123	81,239
	Avg. pkt. enrgy. (in uJ)	10.75	10.99	11.10	13.25	13.65	13.68	11.59	12.01	12.26	12.35	11.87	12.01

Four number of layers — Benchmarks

	263ENC-MP3DEC			MWD			MPEG4			VOPD		
Th.	0.78	0.78	0.68	0.75	0.75	0.72	0.62	0.61	0.58	0.73	0.67	0.65
Lat.(cycle)	78,461	77,450	77,643	99,259	98,262	98,504	96,942	96,261	97,214	92,856	92,342	93,555
Avg. pkt. enrgy. (in uJ)	10.73	10.93	13.21	9.84	9.84	10.68	9.54	10.76	12.41	11.24	11.38	12.95

Benchmarks

	DVOPD			G17			G18			G19		
Th.	0.68	0.67	0.62	0.65	0.63	0.60	0.64	0.63	0.59	0.70	0.67	0.65
Lat.(cycle)	77,961	77,485	77,846	98,959	98,561	98,624	97,242	96,253	97,114	94,356	92,356	93,558
Avg. pkt. enrgy. (in uJ)	9.73	9.83	11.22	9.84	9.84	10.66	7.84	10.97	11.25	11.42	11.10	11.98

Benchmarks

	G20			G21			G22			G25		
Th.	0.72	0.70	0.68	0.82	0.80	0.75	0.65	0.64	0.60	0.79	0.76	0.71
Lat.(cycle)	78,466	77,410	77,856	96,252	95,265	95,305	95,245	94,261	95,014	72,456	71,340	72,550
Avg. pkt. enrgy. (in uJ)	8.75	8.89	10.11	11.04	11.85	11.96	10.54	10.67	10.21	9.24	9.29	9.98

Benchmarks

	G26			G27			G28			G29		
Th.	0.75	0.73	0.71	0.65	0.62	0.59	0.69	0.67	0.59	0.72	0.71	0.69
Lat.(cycle)	87,461	86,480	87,843	83,259	82,260	83,304	87,242	86,260	87,014	84,356	83,340	83,550
Avg. pkt. enrgy. (in uJ)	9.73	9.83	11.22	11.04	11.54	12.06	10.54	10.97	11.21	11.24	11.29	11.98

$\triangle$=Elevator-first and $\triangle'$=Modified Elevator-first

proposed approach of intelligent TSV placement and mapping produces results close to that for the situation with 100% TSVs.

4   Conclusion

This chapter presents techniques for designing 3D-mesh-based NoC systems using single and multiple PSOs. The results produced by single PSO onto mesh are not inspiring. In some cases, the results are not better than our proposed constructive mapping technique. The mapping technique using enriched PSO, that is, augmentations involving multiple PSO, and deterministic initial population, shows reasonable improvement in communication cost while considering the static operation of the system. There is also improvement in dynamic performance and energy consumption of the solutions produced by this strategy, compared to the best ones previously reported as well as our proposed techniques in preceding chapters. It can be noted from the simulation results that, this mapping strategy shows better improvement for the NoCs having a higher number of tasks.

Algorithms which minimize communication cost of the mapping may not consider the thermal effects, resulting in hotspot and high peak temperature. As the NoC consists of different cores, each having its own power profile, area, frequency of operation, etc., it may result in non-uniform heating of the chip. This, in turn, may result in delay variation across the chip. Excessive heating may cause the creation of thermal hotspots. Thus, placement of tasks should be guided not only by their communication requirements but also by their temperature profile. The next chapter focuses on a thermal-aware application mapping strategy to reduce the peak temperature of the die, sacrificing a little communication cost.

References

Bahmani, M., Sheibanyrad, A., Petrot, F., Dubois, F., & Durante, P. (2012). A 3D-NoC router implementation exploiting vertically-partially-connected topologies. In *Proceeding of Computer Society Annual Symposium on VLSI (ISVLSI)* (pp. 9–14). Piscataway: IEEE.

Catania, V., Mineo, A., Monteleone, S., Palesi, M., & Patti, D. (2016). Cycle-accurate network on chip simulation with noxim. *ACM Transactions on Modeling and Computer Simulation, 27*(1), 1–25.

Chen, S., & Montgomery, J. (2011). Selection strategies for initial positions and initial velocities in multi-optima particle swarms. In *Proceeding of Genetic Evolutionary Computation Conference* (pp. 53–60).

Cplex (2013). www.ibm.com/software/in/integration/optimization/cplex

Dubois, F., Sheibanyrad, A., Petrot, F., & Bahmani, M. (2013). Elevator-first: A deadlock-free distributed routing algorithm for vertically partially connected 3D-NoCs. *IEEE Transactions on Computers, 62*(3), 609–615.

Feero, B. S., & Pande, P. P. (2009). Networks-on-Chip in a three dimensional environment: A performance evaluation. *IEEE Transactions on Computers, 58*(1), 32–45.

Luo, G., Zhao, H., & Song, C. (2008). Convergence analysis of a dynamic discrete PSO algorithm. In *Proceeding of International Conference Intelligent Networks and Intelligent System* (pp. 89–92).

Kennedy, J., & Eberhart, R. (1995). Particle swarm optimization. In *Proceeding of International Conference on Neural Networks* (pp. 1942–1948).

Lee, J., & Choi, K. (2013). A deadlock-free routing algorithm requiring no virtual channel on 3D-NoCs with partial vertical connections. In *Proceeding of International Symposium on Networks on Chip (NoCS)* (pp. 1–2). Piscataway: IEEE.

Liu, C., Zhang, L., Han, Y., & Li, X. (2011a). Vertical interconnects squeezing in symmetric 3D mesh Network-on-Chip. In *Proceeding of Asia and South Pacific Design Automation Conference (ASP-DAC)* (pp. 357–362).

Matsumoto, M., & Nishimura, T. (1998). Mersenne twister: A 623-dimensionally equidistributed uniform pseudo-random number generator. *ACM Transactions Model Computer Simulation, 8*(1), 3–30.

Murali, S., & Micheli, G. D. (2004a). Bandwidth constrained mapping of cores onto NoC architectures. In *Proceeding of Design, Automation and Test in Europe (DATE)* (pp 896–901). Piscataway: IEEE.

Rohler, A. B., & Chen, S. (2011). An analysis of sub-swarms in multi-swarm systems. In *Proceeding of Australasian Joint Conference on Artificial Intelligence* (pp. 271–280).

Saito, M., & Matsumoto, M. (2008). SIMD-oriented fast Mersenne twister: A 128-bit pseudorandom number generator. In A. Keller, S. Heinrich, & H. Niederreiter (Eds.), *Monte carlo and quasi-monte carlo methods*. Berlin: Springer.

Tian, X., & Benkrid, K. (2009). Mersenne twister random number generation on FPGA, CPU and GPU. In *Proceeding of NASA/ESA Conference Adaptive Hardware System (AHS)* (pp. 460–464).

Varatkar, G. V., & Marculescu, R. (2004). On-chip traffic modelling and synthesis for MPEG-2 video applications. *IEEE Transactions on Very Large Scale Integration (VLSI) Systems, 12*(1), 108–119.

Wang, P. K., Huang, L., Zhou, C. G., & Pang, W. (2003). Particle swarm optimization for traveling salesman problem. In *Proceeding of International Conference on Machine Learning and Cybernetics* (pp. 1583–585).

Chapter 6
Thermal-Aware Application Mapping Strategy for Designing a 2D NoC-Based Multi-Core Systems

In previous chapters, we have seen a number of application mapping algorithms together with TSV placement to minimize communication cost and energy consumption of NoC-based systems. However, algorithms which minimize communication cost (network latency) of the mapping may not consider thermal effects, resulting in hotspots and high peak temperatures, which in turn decrease the performance of systems, lifetime, reliability and leakage power dissipation. It may also create a very high-temperature variance within the chip, resulting in uneven delays across the chip. The overall problem addressed in this chapter can be stated as follows:

Given the properties of an application (in terms of its task graph) and NoC architecture (in terms of topology graph), an optimum association between tasks and cores has to be determined such that the weighted communication cost (BW × hop-count) and the peak temperature of the die are minimized under a given routing technique.

The inputs to the problem are as follows:

- A task graph G representing the application.
- A topology graph T corresponding to the 2D NoC.
- Power profile of each task, when assigned to a core.
- Power consumption for each router (r_i) and link (l_k).
- A floorplan for the NoC, represented as F.

In this work, it has been assumed that the application will be mapped onto a homogeneous-tile oriented regular 2D mesh-based NoC. A core and its corresponding router together form a tile. It is also assumed that each task has an associated power profile which is the peak power consumed by the task when assigned to a tile.

A PSO-based thermal-aware mapping strategy for 2D NoC has been developed to minimize both communication cost and (peak) temperature of the die for a given application. Furthermore, to check the optimality of the proposed PSO-based

© Springer Nature Switzerland AG 2020
K. Manna, J. Mathew, *Design and Test Strategies for 2D/3D Integration for NoC-based Multicore Architectures*, https://doi.org/10.1007/978-3-030-31310-4_6

approach, an exact method has been developed using ILP technique. We have also compared the proposed work with the other contemporary techniques available in the literature.

1   Temperature Calculation

The temperature of a tile/IP-block depends on its power consumption and its position in the floorplan. Typically, circuits are packaged with a configuration such that the die can be put against the heat spreader. Such a model has been presented in Fig. 6.1. Here, the spreader can be placed against the heat sink which can be cooled by a fan. Considering such configuration, *HotSpot* (Skadron et al. 2003), an efficient thermal modeling tool has been developed to measure the thermal effect at IP-block level. The simple compact model calculates the temperature of each IP-block by considering the heat dissipation within the block and also the effect of heat transfer among the IP-blocks, based on the RC model. The RC model considers three verticals conductive layers for the die, heat spreader and heat sink. It also takes care of the fourth vertical convective layer for the sink-to-air interface. Moreover, it takes into account the heat flow path from the four sides (e.g., north, south, east and west) of spreader, inner and outer portions of the heat sink. Thus, along with the

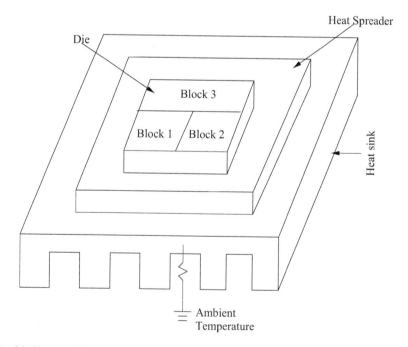

Fig. 6.1 Hotspot (Skadron et al. 2003) model with three layers: die, heat spreader and heat sink

tiles in the floorplan, extra 12 nodes have been considered. Now, the fundamental idea is that the thermal resistance $R_{i,j}^{th}$ of the IP-block IP_i with respect to IP_j can be defined as

$$R_{i,j}^{th} = \Delta T_{i,j}/\Delta P_j \qquad (6.1)$$

where $\Delta T_{i,j}$ represents increment in temperature in IP_i due to unit power dissipated at IP_j and ΔP_j is the unit power dissipation at IP_j. The total number of node in the floorplan is denoted by q. Therefore, thermal resistance matrix can be defined as

$$R^{th} = \begin{pmatrix} R_{1,1}^{th} & R_{1,2}^{th} & \cdots & R_{1,q}^{th} \\ R_{2,1}^{th} & R_{2,2}^{th} & \cdots & R_{2,q}^{th} \\ \vdots & \vdots & \ddots & \vdots \\ R_{q,1}^{th} & R_{q,2}^{th} & \cdots & R_{q,q}^{th} \end{pmatrix} \qquad (6.2)$$

For a given set of power values of individual tiles in NoC, the temperature of each tile can be computed as

$$\begin{pmatrix} T_1 \\ T_2 \\ \vdots \\ T_q \end{pmatrix} = \begin{pmatrix} R_{1,1}^{th} & R_{1,2}^{th} & \cdots & R_{1,q}^{th} \\ R_{2,1}^{th} & R_{2,2}^{th} & \cdots & R_{2,q}^{th} \\ \vdots & \vdots & \ddots & \vdots \\ R_{q,1}^{th} & R_{q,2}^{th} & \cdots & R_{q,q}^{th} \end{pmatrix} \times \begin{pmatrix} P_1 \\ P_2 \\ \vdots \\ P_q \end{pmatrix} \qquad (6.3)$$

where P_i and T_i denotes the power and temperature values for the ith IP-block, IP_i. Thermal resistance matrix can be obtained from *HotSpot* tool for a particular floorplan. The peak temperature of the die can be calculated as

$$T_{peak} = \max\{T_1, T_2, \ldots, T_q\} \qquad (6.4)$$

In this work, we have extracted such thermal resistance matrix using the tool, *HotSpot*, for a given floorplan of an application.

2 ILP Formulation for Thermal-Aware Mapping

An ILP formulation for thermal-aware application mapping problem has been presented in this section. Parameters and variables used for such ILP formulation have been described here. The objective function and constraints have also been described in this section. The parameters and variables used in the ILP formulation are noted in Table 6.1.

Table 6.1 Variables used in ILP formulation

Variables	Definitions
$m_{c_i}^{\mu_s}$	= 1, if ith task c_i is mapped to sth core/tile μ_s
	= 0, otherwise
$P_l^{\mu_s \mu_t}$	= 1, if communication path exists between tiles μ_s and μ_t for a mapped edge l of the task graph
	= 0, otherwise
$D^{\mu_s \mu_t}$	Pre-computed Manhattan distance (MDist) between tiles μ_s and μ_t
Bw^l	Bandwidth requirement of the edge l of the task graph.
$R_{i,j}^{th}$	Thermal resistance value of ith row and jth column of the thermal resistance matrix.
n	Total number of tiles in the NoC.
NT	Set of tiles in the NoC.
q	Total number of units in the floorplan (i.e., $4 \times n + 12$; Skadron et al. 2003)
TU	Set of units in the floorplan.
$P_{q \times 1}$	Pre-computed power value for each units in the floorplan.
T_{peak}	Peak temperature of the die.

2.1   Objective Function

The objective of this work is to minimize the weighted sum of communication cost and peak temperature of the die. The objective function can be expressed as

$$\text{Minimize}: W \times \frac{\left[\sum_{\forall l \in E} BW_l \times \left(\sum_{\forall (\mu_s, \mu_t) \in U} D^{\mu_s \mu_t} \times P_l^{\mu_s \mu_t} \right) \right]}{\beta}$$

$$+ (1 - W) \times \frac{T_{peak}}{\gamma} \tag{6.5}$$

where, E and U represent the set of all edges and tiles respectively. As the communication cost and temperature have different units, normalization of these two metrics is approximated by assuming a worst-case scenario. More precisely, β is set to $\sum_{\forall l \in E} (BW_l \times 2n)$ for an $n \times n$ NoC, where $2n$ is the diameter of the NoC. Parameter γ is fixed to peak temperature of mapping obtained by sorting the tasks in decreasing order of power values and mapping consecutive cores in this list at close proximity. W is a weight factor (between 0 and 1) meant to balance the communication cost and peak temperature of the die. When $W = 0$, it minimizes the die temperature, and when $W = 1$, it considers only communication cost minimization.

2.2 Constraints

The following set of constraints have been framed to solve the thermal-aware application mapping problem:

1. **Mapping constraints**

$$\forall c_i \in C, \ \sum_{\mu_s \in U} m_{c_i}^{\mu_s} = 1 \tag{6.6}$$

$$\forall \mu_s \in U, \ \sum_{c_i \in C} m_{c_i}^{\mu_s} \leq 1 \tag{6.7}$$

In this work, it has been considered that any task is associated with single tile. Expression (6.6) ensures this condition. However, it has also been considered that each tile can be associated with at most one core. To guarantee this condition, inequality (6.7) has been introduced.

2. **Path constraint for communication**
Each edge (l) in task graph, i.e., two communicating tasks such as c_i and c_j, must be mapped onto two different tiles and a communication path ($P_l^{\mu_s \mu_t}$) between those two tiles ($\mu_s \mu_t$) must be created based on the routing algorithm. It can be ensured by the following constraint:

$$\forall l \in E, \forall (\mu_s, \mu_t) \in U, \ P_l^{\mu_s \mu_t} = \begin{cases} 1, & \text{if } (m_{c_i}^{\mu_s} = 1) \text{ and } (m_{c_j}^{\mu_t} = 1) \\ 0, & \text{Otherwise} \end{cases} \tag{6.8}$$

Moreover, constraints (a) and (b) have been imposed for an ILP formulation of the above constraint.

(a) $\forall l \in E, \forall (\mu_s, \mu_t) \in U \ [P_l^{\mu_s \mu_t} \geq m_{c_i}^{\mu_s} + m_{c_j}^{\mu_t} - 1]$, where c_i and c_j are the source and destination tasks for the lth edge respectively which are associated with cores μ_s and μ_t respectively.
(b) $\forall l \in E, \forall (\mu_s, \mu_t) \in U \ [P_l^{\mu_s \mu_t} \leq (m_{c_i}^{\mu_s} + m_{c_j}^{\mu_t}) \div 2]$, where c_i is source of lth edge and it is mapped to tile μ_s and c_j is destination of lth edge and its mapped to tile μ_t.

3. **Constraint to order the power value according to mapping**

$$\forall i \in TU, \ P_{i \times 1} = \begin{cases} \sum_{j=1}^{n} K_j \times m_{c_j}^{r_i}, & \text{if } i \leq n \\ K_i, & \text{Otherwise} \end{cases} \tag{6.9}$$

where, power consumed by the ith unit in the floorplan is denoted as P_i.

According to the mapping, the association between tasks and tiles can be changed which in turn changes the order of power values in power matrix, P. The above constraint guarantees that ordering.

4. **Temperature constraint**

The temperature of individual IP-blocks in the floorplan can be calculated as the product of the power consumption of that block and the corresponding thermal resistance. Moreover, the peak temperature (T_{peak}) of the die can be found using the following constraint.

$$\forall i \in NT, \ T_{peak} \geq \sum_{j=1}^{q} r_{ij} \times P_j \tag{6.10}$$

The above formulated ILP has been solved using CPLEX (Cplex 2013) to get the optimum solution. In the following, a Particle Swarm Optimization (PSO) based technique has been proposed including several augmentations to find the solution for the bigger size of NoC-based systems.

3 PSO Formulation for Thermal-Aware Mapping

In this work, the DPSO technique has been used similar to the last chapter. However, the particle formulation and fitness calculation have been modified according to the problem. The initial particle generation process of PSO technique includes a thermal-aware and a communication-aware mapping strategy to generate some intelligent particles.

3.1 Particle Formulation and Fitness Calculation

3.1.1 Particle Structure

A particle for the thermal-aware mapping problem corresponds to an association between tasks and tiles. For instance, a mapping of 9 cores on to 3 × 3 router grid is shown in Fig. 6.2a. Here, r and c denote the tiles and tasks in the NoC, respectively.

In this work, it has been assumed that tiles are arranged in row-wise manner and numbers are assigned from top left to bottom right corner in a 2D layout in an increasing order. So, a particle can efficiently be represented as 1D-array, in which indices represent the tile number and the value within the cell represents the task associated with the tile. A particle is a permutation of task numbers, identifying task associated with the tiles in the NoC topology. Figure 6.2b presents the particle structure corresponding to the NoC mapping in Fig. 6.2a. The fitness of each particle is evaluated using expression (6.11), which is the sum of weighted communication

Fig. 6.2 (**a**) 3 × 3 NoC
topology. (**b**) Corresponding
particle structure

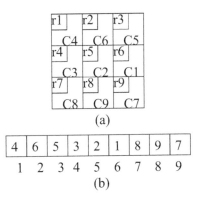

(a)

4	6	5	3	2	1	8	9	7
1	2	3	4	5	6	7	8	9

(b)

cost and peak temperature of the die which depends upon the position of cores in
the particle.

$$Fitness = W \times \frac{CommCost}{\beta} + (1 - W) \times \frac{T_{peak}}{\gamma} \qquad (6.11)$$

where weight is denoted by W. The $CommCost$, T_{peak}, β and γ have been defined
earlier in Sect. 2.1.

The particles evolve based on the experience of their local best values and the
global best of the current generation. The local best of a particle is decided in a
similar way, as discussed in the last chapter. Similarly, the global best particle is
selected from the set of local best values.

3.2   Augmentation to the Basic PSO

This section describes the augmentations included in the basic PSO technique. It
incorporates all the techniques discussed in Chap. 5 together with the modified
initial population generation discussed next.

3.2.1   Initial Population Generation

To guide the exploration, a few particles have been generated using the following
strategies. These particles are tuned towards giving solutions with either better
communication cost or with reduced temperature.

(a) **Communication-aware**: To generate some particles which can perform well
    in terms of communication cost, this work has used the deterministic initial
    population generation strategy similar to Chap. 4. The technique can generate
    a number of mapping candidates equal to the number of tiles in the NoC,

quite fast. In this technique, a task sequence is generated by sorting the communication edges in decreasing order. The first task, in such order, can be mapped to any of the n tile positions of an NoC with n tiles while the remaining tasks are to be mapped into close proximity based on outgoing edges of the already mapped tasks.

(a) **Thermal-aware**: To improve the solution quality in terms of thermal behaviour, a few particles in the initial generation have been created using a deterministic thermal-aware mapping technique, discussed here. Temperature of a tile depends on its power consumption as well as its location in the floorplan. For a NoC topology with n tiles, we include n deterministically created particles in the initial population.

First, the tasks are sorted in decreasing order of their power consumption values. This ordering of tasks is recorded in CP_{seq}. Next, we generate n unique sequences, n being the number of tasks in the application. Each sequence starts with a different task. The sequence corresponding to the task, $task\_id$ is stored in $M_{seq}[task\_id]$. The first entry in this sequence is the task itself. The next entry is the task having the maximum communication with task $task\_id$. Successive tasks are attached to the sequence, selected on the decreasing order of their communication with the previously placed tasks in the sequence. The process has been described in procedure $Create\_seqence$.

---

procedure CREATE_SEQUENCE(G)   ▷ G: Application task graph
    for each task $task_{id}$, in G **do**
        Unmark all tasks in G
        $M_{seq} \leftarrow task_{id}$
        Mark $task_{id}$
        while there are unmarked tasks in G **do**
            Select unmarked task (say $task_{max}$) having maximum
            communication with marked tasks
            Append $task_{max}$ to $M_{seq}[task_{id}]$
            Mark $task_{max}$
        end while
    end for
end procedure

Now, a distance for ith sequence($M_{seq}[i]$), gap_i, is computed with respect to CP_{seq} as

$$gap_i = \sum_{j=1}^{n} |position(M_{seq}[i], j) - position(CP_{seq}, j)| \qquad (6.12)$$

where $position(s, t)$ is index of tth task in sequence, s. Actually, gap_i quantifies the uniqueness of ith sequence, $M_{seq}[i]$, with respect to sequence, CP_{seq}. The quantity increases as similarity decreases between two sequences. The upper and lower bound of gap can be found, as if, both the sequences are completely opposite to each other or same, respectively. However, both these sequences are unacceptable for mapping. A sequence with mix of high and low power consuming tasks is expected to be beneficial for both communication cost and temperature minimization. Such sequence has the average/almost average gap value. The strategy has been explained in procedure *Find_avg_sequence*.

---

procedure FIND_AVG_SEQUENCE(CP_{seq}, M_{seq})
//CP_{seq}: Task sequence in decreasing power consumption, M_{seq}: Set of task order of sequences

    $Avg \leftarrow 0$
    for each sequence k in M_{seq} **do**
        $gap_k \leftarrow 0$
        for each task, c_i, in sequence $M_{seq}[k]$ **do**
            $gap_k \leftarrow gap_k + |position(M_{seq}[k], c_i) - position(CP_{seq}, c_i)|$
        end for
        $Avg \leftarrow Avg + (gap_k/n)$
    end for
    Let m be the sequence, such that, $|Avg - gap_m/n|$ is the minimum
    return m
end procedure

After finding the average sequence, procedure, *find_mapping* is called to obtain n different mappings by putting the first task in the sequence at find each tile position. The *find_mapping* process works as follows.

It takes starting task and tile position to start the mapping. It maps the first task (say, c_1) to the provided tile position (say, μ_1). Assume that c_1 has been mapped onto tile μ_1 in the NoC topology. To map the next task from the sequence $M_{seq}[start\_task]$, it finds the neighbouring (one-hop away) tiles from μ_1. Suppose, μ_1 has neighbours μ_2, μ_3 and μ_4. Since, all these tiles are one-hop away from μ_1, all are equally suitable for the mapping of next task (say c_2). In general, at any instance, a subset of tasks are already mapped onto the tiles of the topology graph. Let, C' and R' be the set of already mapped tasks and the set of corresponding tiles, respectively. The algorithm, now, selects the next task from the sequence $M_{seq}[start\_task]$. Let, the selected task be c_i. Then, it tries out mapping of c_i to each tile location one-hop away from any tile in R'. The communication cost is calculated for each individual mapping candidate, considering the sub-graph consisting of tasks in the set $C' \bigcup \{c_i\}$. If there is a unique mapping candidate with minimum communication cost, it is accepted for mapping of c_i and the process continues with next selected task from the sequence, $M_{seq}[start\_task]$, in a similar fashion. However, if several mapping candidates of same cost exist for c_i, suppose $P = \{p_1, p_2, \ldots, p_k\}$ be the set of k candidate positions. Among all these k positions, let, algorithm choose p_1, for the time being, to be the mapping of c_i. It continues to map the remaining tasks in same way as mentioned earlier. At this moment, the algorithm does not distinguish between contending positions with minimum cost value. Rather, it chooses the first such position and carries on with unmapped tasks. The predicted cost is calculated when all tasks in the task graph are mapped by considering selected tile position p_1 for c_i. For other $k - 1$ positions, such as, $p_2, p_3, \ldots, p_k$, costs are calculated in a similar way and task c_i is associated with the tile position with the minimum predicted cost. The process continues by selecting the tasks from the same sequence.

---

procedure FIND_MAPPING(R_{pos}, S_{task}, G, R)
//G and R task and topology graphs, R_{pos}: Tile number in the R, S_{core}: Task to be mapped into tile R_{pos}
    Map S_{task} to tile R_{pos}
    Mark S_{task} as mapped
    for next unmarked task, c, in sequence, $M_{seq}[S_{task}]$ **do**
        Compute cost for all one-hop mappings of c to unmapped tiles
        Find tile locations with same minimum costs in L_r
        for each location p in L_r **do**
            Attach c with p
            Map all unmapped tasks from $M_{seq}[S_{task}]$ to any tile which are one-hop away from the already mapped one with minimum cost.
            Compute cost of the such mapping.
            Map c to any location in L_r with minimum cost
        end for
        Identify the mapping with minimum cost
    end for
    return mapping with minimum cost
end procedure

3.3 PSO-Based Thermal-Aware Mapping Technique

The entire PSO engine has been described in Algorithm 6.1. In this algorithm, each *Particle* represents a configuration identifying the mapping of tasks to the tiles of NoC topology. Such configuration represents a position in the solution space. The algorithm generates few intelligent and remaining random particles, making a total of $NPart$ number of particles in the initial generation of PSO (line 10). The few intelligent configurations—communication and thermal-aware are generated by using the strategy, described earlier. The local best of each particle is set to itself. The global best particle is set to be the best configuration found in the local best configuration set (line 21). Each configuration is judged based on expression (6.11). Here, *NPart* is the number of particles while *MGEN* is the maximum number of generations. The number of PSO runs is represented by *MPSO*.

After generating the initial configurations and finding the global best configuration at first generation, the particles are evolved using *UpdatePart()*. Each particle evolves by sharing the experience of its local as well as global best of the generation with a random probability. The *swap* sequences are generated based on the particle, local and global best configurations (as discussed in the last chapter). New particle or configuration is generated by applying those swap sequences onto the particle

---

Algorithm 6.1 PSO-based thermal-aware mapping

Input: Task graph C, Topology graph T, Floorplan of the NoC and Resistance matrix, Power profile of the tasks, router and links

Output: Mapping of C to T

1: Set $MGEN$, $MPSO$ and $NPart$
2: Set W
3: $BestFitness \leftarrow \infty$
4: **for** m from 0 to $MPSO$ **do**
5:      $IM \leftarrow true$
6:      $BeforeIMBestFitness \leftarrow \infty$
7:      **while** $IM = true$ **do**
8:          $gen \leftarrow 0$
9:          **while** $gen < MGEN$ and $IM = true$ **do**
10:              **for** p from 0 to $NPart$ **do**
11:                  **if** $gen = 0$ and $m = 0$ **then**//Initialization
12:                      $Particle_p \leftarrow \{\text{ThermIntelligent}(), \text{CommIntelligent}(), \text{Random}()\}$
13:                      $Particle_p^{pbest} \leftarrow Particle_p$
14:                  **else**//Update particle
15:                      UpdatePart($Particle_p$,$Particle_p^{pbest}$, $Particle^{gbest}$, $ProbRandom_{pbest}$, $ProbRandom_{gbest}$)
16:                      Compute fitness of $Particle_p$ using expression (6.11)
17:                      UpdateLBest($Particle_p^{pbest}$, $Particle_p$)
18:                  **end if**
19:              **end for**
20:              Compute global best particle $Particle^{gbest}$ from the set of particles $Particle^{pbest}$
21:              Update the $BestFitness$ as global best as per the requirement and reset the gen counter
22:              **if** $BeforeIMBestFitness = BestFitness$ **then**
23:                  $IM \leftarrow false$
24:              **end if**
25:              $gen \leftarrow gen + 1$
26:          **end while**

27:          **if** $IM = true$ **then**
28:              $BeforeIMBestFitness \leftarrow BestFitness$
29:              Perform Inverse Mutation (IM) at random position of each particle of the last generation
30:          **end if**
31:      **end while**
32:      **for** p from 0 to $NPart$ **do**
33:          $Particle_p \leftarrow \{\text{Random}()\}$
34:          CopyLbest($Particle_p^{pbest}$)
35:      **end for**
36: **end for**

---

with a random probability. Furthermore, the local best of each particle gets changed by comparing the fitness between its current *lbest* and the new particle (line 17). After each generation, the PSO engine checks the $BestFitness$ value. If the best fitness does not improve, PSO keeps count of the generations. The *inversion mutation* (IM) (line 29) can be applied onto each particle, if generation count reaches up to *MGEN* and *IM* is *true*. Furthermore, the PSO engine can create new particles by assigning random configurations, whereas, local best values are passed from the earlier PSO run when the generation count reaches to its maximum value and IM is *false* (line 32 to 35). The PSO engine stops when multiple PSO count reaches to a maximum value (*MPSO*).

4 Experimental Results and Analysis

This section presents our experimental results on different NoC benchmarks and compare these with ILP and other existing techniques for a number of NoC benchmark applications. The simulation environment is same as that in Sect. 5. The proposed scheme also has been evaluated, using highly scalable benchmarks, such as Big data and Graph analytical workloads (see Sect. 4.7). In this work, a fixed ambient temperature of 45 °C has been assumed (Mahajan 2002). The power consumption for NoC components, such as routers and links, has been calculated using *Orion* (Kahng et al. 2012), and the core power values by using *McPAT* (Li et al. 2013) for the *Alpha* 21,264 cores, with 45 nm technology node. For the simulation of the present work, a 2D mesh-based NoC with grid size of 1 mm and 45 nm technology node, has been considered as in Zhu et al. (2015). The power consumption of NoC components, as also that of the core, forms the tile power. The applications have been mapped onto a 2D mesh-based-NoC with mesh sizes shown in Table 6.2. The results are organized as follows. First, to check the efficiency of PSO formulation, the results have been compared with those of the ILP-based approach for different weight factors, W. The results corresponding to different augmentations to the basic PSO have been noted next. This has been followed by comparing the dynamic performances and energy consumptions of the solutions.

4.1 Comparison Across Mapping Techniques

For checking the quality of the proposed PSO approach for thermal-aware mapping, the results of PSO are compared with those of ILP, TAPP (Zhu et al. 2015) and CoolMap (Moazzen et al. 2012) (see Table 6.3). The results show that the solution obtained by PSO for applications PIP, MPEG4 and VOPD is the same as the one reported for ILP. However, for other applications, ILP either could not start or get terminated due to the creation of numerous constraints. Actually, the number of variables needed in ILP formulation, as mentioned in Sect. 1, is proportional to the

NoCs	No. of Cores	2D Mesh
PIP	8	4 × 2
263ENC-MP3DEC	12	3 × 4
MWD	12	3 × 4
MPEG-4	12	3 × 4
VOPD	16	4 × 4
DVOPD	32	4 × 8
G17	64	8 × 8
G18	64	8 × 8
G19	64	8 × 8
G20	64	8 × 8
G21	64	8 × 8
G22	64	8 × 8
G28	128	8 × 16
G29	128	8 × 16
G30	128	8 × 16

Table 6.2 Benchmark applications and their mesh sizes

square of the product of the number of threads in the task graph and the number of cores in the topology graph. Table 6.3 also shows the CPU time taken by each mapping method. In CoolMap, the weight factor, W, has not been considered, as it only provides the result for $W = 0$, i.e., thermal-aware mapping. PSO could produce satisfactory results within reasonable time. A commercial ILP solver, CPLEX (Cplex 2013), has been used to solve the formulated ILP. Both ILP and PSO have been implemented on a Dell PowerEdge T410 system with 8 cores (Intel Xeon processor, E5606@2.12 GHz), 64 GB RAM. The capacity of the PSO to achieve the optimal results, found from ILP, corroborates the quality of the PSO.

4.2 Effect of Inversion Mutation (IM) and Randomness on the Augmentation of Basic PSO

The effect of incorporating IM and SMIT random number generator into the basic PSO, on PSO's augmentation, is discussed in this section. The results corresponding to this incorporation are shown in Table 6.4 for applications G28 and G29. The second column of the table shows the results of the basic PSO without any augmentation, while the third and fourth columns show the results of incorporating inverse mutation and SIMT into this basic PSO. It can be seen from the table that better results have been achieved after incorporating the IM and the SIMT random number generator. The results reported here are for $W = 1$, that is, communication-aware mapping. This establishes the suitability of the augmentation strategies proposed here for improving the solution quality.

Table 6.3 Comparison between ILP, DPSO, CoolMap (Moazzen et al. 2012) and TAPP (Zhu et al. 2015) approaches, in terms of communication cost, peak temperature and CPU time, with variable weightage, W

| NoCs | W | Thermal-aware mapping techniques | | | | | | | |
| | | ILP | | PSO | | CoolMap (Moazzen et al. 2012) | | TAPP (Zhu et al. 2015) | |
		CC, T_{peak} (°C)	CPU time (s)	CC, T_{peak}(°C)	CPU time (s)	CC, T_{peak} (°C)	CPU time (s)	CC, T_{peak} (°C)	CPU time (s)
PIP (8 cores)	0	1216, 72.26	0.12	1216, 72.26	0.02	1448, 84.40	0.015	1400, 76.12	0.018
	0.5	640, 83.75	0.17	640, 83.75	0.02	-,-	-	640, 83.75	0.018
	1	640, 83.75	0.17	640, 83.75	0.01	-,-	-	640, 83.75	0.017
MWD (12 cores)	0	1986, 73.14	0.11	1986, 73.14	0.03	2438, 82.56	0.026	2385, 79.93	0.027
	0.5	1258, 82.73	0.15	1258, 82.73	0.03	-,-	-	1348, 83.73	0.027
	1	1248, 80.51	6.32	1248, 80.51	0.02	-,-	-	1248, 85.23	0.028
MPEG4 (12 cores)	0	7449.50, 69.69	0.11	7449.50, 69.69	0.03	10,123.61, 74.23	0.024	9025.23, 72.12	0.028
	0.5	3643, 73.29	0.52	3643, 73.29	0.03	-,-	-	4831, 78.12	0.027
	1	3587, 82.53	7.45	3587, 82.53	0.01	-,-	-	3632, 85.35	0.028
263ENC-MP3DEC (12 cores)	0	551.61, 71.15	0.11	551.61, 71.15	0.03	732.32, 76.23	0.027	638.12, 75.12	0.028
	0.5	310.94, 76.81	0.15	310.94, 76.81	0.03	-,-	-	310.12, 77.31	0.027
	1	230.43, 84.51	6.32	230.43, 84.51	0.02	-,-	-	230.43, 80.51	0.027

Table 6.4 Comparison of different augmentation techniques, in terms of communication cost

NoCs	Basic PSO	Basic PSO with IM only	Basic PSO with SIMT only
G28	574,527.63	54,888.44	569,070.06
G29	435,423.03	413,501.22	414,277.16

4.3 Comparison with Other Methods

This section compares the experimental results of the proposed approach with those of several recent approaches reported in the literature (see Table 6.5). The columns marked as 'Single PSO' and 'Multiple PSO' correspond to the cases in which PSO has been run only once, or several times, as mentioned in Sect. 3.2. Furthermore, the column 'Random initial population', under 'single PSO', represents the situation in which the augmentations suggested in Sect. 3.2 have not been incorporated. On the other hand, the columns marked as 'Augmentation' correspond to the results relating to all the augmentations used. From Table 6.5, it can be observed that the proposed augmented multi-PSO-based approach performs better than other contemporary thermal-aware approaches reported in the literature when $W = 0$. The methods have been ranked, based on their temperature profile only, because only temperature has been optimized in this experiment. Multiple PSO results have been taken as unity. The temperature profile generated by CoolMap and TAPP are, on the average, 15% and 11% away, respectively, from the proposed method.

The CPU times for the state-of-the-art methods and the proposed strategy are shown in Table 6.6. It can be seen from this table that the proposed method, in comparison to the non-PSO-based methods, requires a little longer CPU time. More precisely, considering the 'CoolMap' timing results as unity, the proposed approach and 'TAPP' method require, on the average, about 65% and 25% more CPU time. However, the proposed approach produced better quality solutions than those of all other state-of-the-art techniques: 'CoolMap' and 'TAPP'.

4.4 Dynamic Performance Comparison

For better comparison with the thermal-aware mapping solutions, the resulting NoCs have been simulated, using *Noxim* (Vincenzo et al. 2016), a cycle-accurate simulator. Here, the cores send messages by following the traffic pattern in accordance with the edge weights of the task graph. However, as mentioned in Sahu et al. (2014b), the time distribution of the messages follows self-similar nature. Each core generates traffic in a self-similar fashion by aggregating a large number of ON-OFF message sources following Pareto distribution with Hurst parameter, $H = 0.75$ and Shape parameters $\alpha_{ON} = 1.5$ and $\alpha_{OFF} = 1.17$ (Sahu et al. 2014b). The configuration of the Noxim simulator has been presented in Table 6.8. Any NoC design is expected to have high throughput, while the average latency is expected to

Table 6.5 Comparison of the existing methods, in terms of communication cost and peak temperature

NoCs	CoolMap (Moazzen et al. 2012) CC, T_{peak}(°C)	TAPP (Zhu et al. 2015) CC, T_{peak}(°C)	Proposed PSO-based approaches ($W = 0$)						
			Single PSO			Multiple PSO			
			Random initial population CC, T_{peak}(°C)	Augmentation		CC, T_{peak}(°C)	Augmentation		
				Comm.-aware CC, T_{peak}(°C)	Comm. and Temp.-aware CC, T_{peak}(°C)		Comm.-aware CC, T_{peak}(°C)	Comm. and Temp.-aware CC, T_{peak}(°C)	
PIP	1448, 84.4	1400, 76.12	1432.00, 76.63	1391.00, 76.23	1388.00, 73.23	1365.00, 77.23	1263.00, 75.78	1222.00, 72.26	
MWD	2438, 82.56	2385, 79.93	2402.00, 81.34	2232.00, 83.76	2172.00, 78.43	2027.00, 78.12	1996.00, 76.15	1982.00, 73.14	
MPEG4	10,123.61, 74.23	9025.23, 72.12	9127.42, 74.52	8125.23, 76.16	8110.42, 73.21	8138.50, 71.56	7462.50, 70.53	7334.51, 69.69	
263ENC-MP3DEC	732.32, 76.23	638.12, 75.12	565.13, 79.18	559.48, 82.21	552.23, 74.24	562.76, 73.11	555.34, 73.51	527.61, 71.15	
VOPD	15,391.00, 83.13	15,258.00, 80.51	16,230.00, 85.98	10,144.00, 85.92	9995.00, 80.89	7162.00, 78.69	4965.00, 76.89	4899.00, 72.17	
DVOPD	18,167.00, 85.53	16,749.00, 82.28	17,457.00, 84.16	15,682.00, 83.07	15,662.00, 79.27	13,480.00, 79.43	11,070.00, 76.16	10,482.00, 71.07	
G17	100,320.00, 87.20	99,660.32, 83.32	108,983.12, 85.91	105,556.14, 84.15	102,367.43, 79.23	98,144.35, 83.11	97,375.23, 80.18	97,115.76, 78.43	
G18	27,623.42, 86.78	25,355.76, 84.12	23,573.44, 86.13	23,452.13, 87.19	23,433.23, 78.15	23,139.62, 80.11	22,876.34, 78.82	22,127.79, 74.53	
G19	27,099.91, 87.14	24,637.14, 85.22	24,162.16, 86.26	23,743.16, 85.66	23,635.13, 80.65	22,135.49, 83.87	20,574.14, 79.75	20,128.13, 76.31	
G20	348,658.13, 85.10	329,274.15, 82.21	332,143.17, 86.18	301,246.15, 87.18	299,848.43, 78.54	297,239.22, 79.72	285,129.22, 76.45	284,689.00, 71.18	
G21	344,058.32, 88.41	320,569.15, 86.18	361,245.18, 85.14	347,915.14, 84.54	347,897.18, 81.65	313,475.16, 82.14	299,797.13, 79.27	294,657.18, 76.24	
G22	196,587.46, 88.51	152,841.22, 84.55	189,361.12, 87.16	161,025.76, 87.78	159,990.85, 78.51	157,697.53, 80.12	136,798.86, 78.75	114,689.54, 74.61	
G28	995,975.26, 89.17	989,951.19, 87.45	989,983.65, 89.72	987,527.14, 89.19	987,338.92, 83.19	984,366.12, 86.04	957,856.13, 83.17	956,745.14, 79.86	
G29	708,963.34, 88.82	688,960.16, 86.49	681,551.24, 87.43	665,633.35, 86.17	665,439.13, 78.16	689,353.22, 84.14	654,318.19, 79.14	653,261.18, 72.27	
G30	276,583.25, 89.34	257,658.27, 86.22	249,291.00, 88.10	236,796.26, 87.22	228,777.22, 80.21	220,638.72, 84.22	212,168.11, 82.7	201,272.19, 77.16	
Rank (on Temp.)	1.15	1.11	1.14	1.14	1.06	1.09	1.05	1	

Table 6.6 Comparison of CPU time

NoCs	Methods		
	MPSO	CoolMap (Moazzen et al. 2012)	TAPP (Zhu et al. 2015)
	CPU time in s		
G17	10.29	7.12	8.75
G18	22.13	10.56	15.23
G19	27.19	11.21	17.65
G20	23.22	12.29	16.85
G21	15.32	8.23	10.19
G22	17.23	10.34	12.11
G28	792.11	413.23	502.51
G29	864.40	553.12	732.12
G30	670.14	450.32	523.12
Rank	1.65	1.00	1.25

be low (Feero and Pande 2009). The throughput and latency values have respectively been marked as 'Th.' and 'Lat.' in Table 6.7. Considering these values as unity for CoolMap (Moazzen et al. 2012), the throughput of the proposed approach increases by 34.02% and the latency reduces by 11.13%. The corresponding figures for TAPP (Zhu et al. 2015) are 17.94% and 7.19%, respectively (Table 6.8).

4.5 Trading-Off Peak Temperature and Communication Cost

To increase system safety, a trade-off is established between NoC temperature and communication cost, using expression (6.11). The weight factor, W, has been varied from 0 to 1. When $W = 0$, the expression considers only the NoC temperature for optimization, and when $W = 1$, only the communication cost. In this experiment, the varied W values are 0, 0.2, 0.5, 0.8 and 1. The expression has been used as a cost function in the proposed augmented multi-PSO (AMPSO) strategy.

Figures 6.3 and 6.4 show the trade-offs between communication cost and peak temperature for applications VOPD and G29, respectively. The values corresponding to peak temperature and communication cost have been marked in the figures, respectively, as 'PSO-Temp' and 'PSO-CC'. These trade-off values are compared with those of TAPP (Zhu et al. 2015). The values corresponding to communication cost and peak temperature of TAPP have been marked in the figures, respectively as 'TAPP-CC' and 'TAPP-Temp'. From the figures it can be seen that the proposed AMPSO method explores better trade-off than that of TAPP (Zhu et al. 2015). The trend has been found to be the same for other application too. The designer can thus have the liberty to choose the solution that suits the requirement.

Table 6.7 Comparison of the existing methods, in terms of dynamic performance

Mapping techniques	CoolMap (Moazzen et al. 2012)	TAPP (Zhu et al. 2015)	Proposed	CoolMap (Moazzen et al. 2012)	TAPP (Zhu et al. 2015)	Proposed	CoolMap (Moazzen et al. 2012)	TAPP (Zhu et al. 2015)	Proposed
NoCs	PIP			MWD			MPEG-4		
Th.	0.69	0.72	0.78	0.62	0.67	0.73	0.59	0.63	0.67
Lat.	89,475.2	87,369.28	82,134.56	89,978.7	75,264.9	70,189.24	92,679.12	79,998.5	80,169.35
NoCs	263ENC-MP3DEC			VOPD			DVOPD		
Th.	0.79	0.83	0.86	0.59	0.63	0.67	0.55	0.64	0.68
Lat.	81,977.1	78,254.37	75,874.67	82,098.14	79,066.40	78,334.81	83,735.70	79,037.40	76,587.95
NoCs	G17			G18			G19		
Th.	0.45	0.49	0.67	0.38	0.56	0.67	0.70	0.75	0.78
Lat.	103,185	91,149.20	89,535.70	103,845	90,482.10	88,785.40	90,854.10	90,247.40	82,755.70
NoCs	G20			G21			G22		
Th.	0.57	0.60	0.70	0.42	0.60	0.65	0.61	0.65	0.79
Lat.	90,772.20	90,736.30	83,938.10	90,523.40	89,742.67	84,260.40	90,716.70	90,516.20	82,791.20
NoCs	G28			G29			G30		
Th.	0.34	0.43	0.48	0.32	0.51	0.61	0.45	0.53	0.65
Lat.	103,489	91,979	90,165	102,787	90,187.90	89,963	106,757	93,891.90	87,978

Table 6.8 Noxim settings

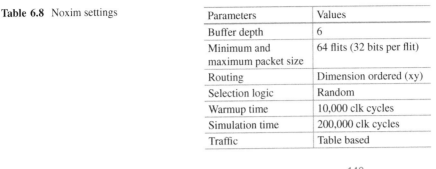

Parameters	Values
Buffer depth	6
Minimum and maximum packet size	64 flits (32 bits per flit)
Routing	Dimension ordered (xy)
Selection logic	Random
Warmup time	10,000 clk cycles
Simulation time	200,000 clk cycles
Traffic	Table based

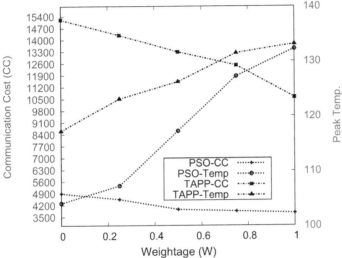

Fig. 6.3 Trade-off between peak temperature and communication cost of application VOPD

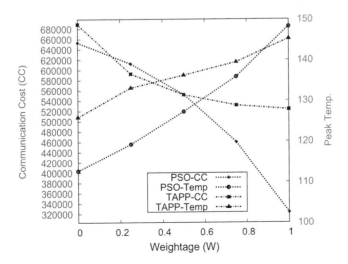

Fig. 6.4 Trade-off between peak temperature and communication cost of application G29

4.6   Imposing Thermal Safety

In this experiment, the PSO formulation has been so modified that, by taking peak temperature as a constraint, it finds out the mapping solution that is most suitable to the temperature budget. If the PSO fails in finding a suitable solution, then it provides the best thermal performing solution. Table 6.9 presents the given temperature constraints (T_{const}), communication costs (CC) and the achievable peak temperatures (T_{peak}). Relaxing the constraint on peak temperature could lead to better solutions (in terms of communication cost) and vice versa.

4.7   Experimentation with Big Data and Graph Analytical Workloads

In this experiment, the proposed strategy has been evaluated for highly scalable benchmarks, *PARSEC* (Bienia et al. 2008). For Big data application, in PARSEC, two inputs—*small* and *large*—are considered here to understand how workload behaviour varies with the size of input data. The PARSEC benchmark suit defines a number of input set size, here, the authors consider <simsmall> and <simlarge>, whereas for application with Graph analytical workload, multi-threaded Breadth First Search (BFS) (Bader and Madduri 2006) has been chosen, with input graphs Kronecker (Kron) and Uniform from GAP benchmarks (Beamer et al. 2015). They behave as Graph 500 benchmarks (gra 2018). Table 6.10 shows the benchmarks and inputs used for the experiments. Traces have been collected from real application, using the full execution-driven simulator SIMICS (Magnusson

Table 6.9 Communication costs and peak temperature for given temperature constraints

Benchmarks					
VOPD		DVOPD		G19	
T_{const}	CC, T_{peak}	T_{const}	CC, T_{peak}	T_{const}	CC, T_{peak}
87	3612, 85.35	85	9427, 83.25	87	14,618.34, 86.38
85	4888, 84.07	82	10,486, 82.10	85	20,342.73, 84.43
82	4899, 78.07	85	10,510, 84.10	85	20,207.73, 83.43

Table 6.10 Benchmarks considered along with input sizes

	Input size	
Benchmarks	Simsmall	Simlarge
Blackscholes	4096 options	65,536 options
Dedup	10 MB	184 MB
Ferret	16 queries	256 queries
Raytrace	480 × 270 pixels	1920 × 1080 pixels

Table 6.11 Cache and core parameters used for PARSEC and GAP application suit simulation

Parameter	Value
L1/L2 coherence	MOESI
L2 cache size/assoc	4 MB/16-way
L2 cache line size	64
L2 access latency (cycles)	4
L1 cache size/assoc	64 MB/4-way
L1 cache line size	64
L1 access latency (cycles)	2
Core frequency (GHz)	5
Threads (core)	1
Issue policy	In-order
Memory size (GB)	4
Memory controllers	16
Memory latency (cycle)	160
Directory latency (cycle)	80

et al. 2002), with the memory packages GEMS (Martin et al. 2005) enabled for 64 cores. The performance on PARSEC workloads has been evaluated. Table 6.11 shows the core and cache parameters used for PARSEC workloads. After collecting the traces, thread-level analysis has been done to capture the communication behaviour. Next, thermal-aware mapping techniques have been performed and the performance measured, in terms of throughput and latency, using Noxim simulator. Table 6.12 compares the performances on PARSEC and GAP benchmarks with different workloads. For larger input or graph size, the performance has been found to degrade, due to more access in the memory unit than the other units, which results, in creating more temperature in the core. The thermal-aware mapping technique attempts to place high-temperature cores far away, and this degrades the performance in comparison to that of smaller input sizes. The proposed PSO-based technique (when $W = 0$) outperforms 'TAPP' and 'CoolMap', because it explores the solution space rigorously, using several augmentation techniques.

5 Conclusion

This chapter presents a technique to design of 2D mesh-based NoC, ensuring thermal safety, using single PSO and multiple PSO. The results produced by using single PSO onto mesh are not that good. However, the enriched PSO, having several augmentations including multiple PSO with deterministic initial population, shows reasonable improvement in communication cost and temperature while considering static operation of the system. To check the optimality of the solution, an ILP-based solution has also been proposed. The dynamic performance of solution produced by PSO also shows the improvement compared to the previously reported works in the literature. It can be noted from the simulation results that, this mapping

Table 6.12 Performance measurement of PARSEC and GAP benchmarks with small and large input size

Benchmarks		Input size					
		Small			Large		
PARSEC	Parameters	PSO	TAPP	CoolMap	PSO	TAPP	CoolMap
Black	Th.	0.68	0.61	0.53	0.56	0.52	0.49
scholes	Lat.	83,142	85,431	87,219	89,321	91,538	93,418
Dedup	Th.	0.72	0.65	0.61	0.65	0.61	0.59
	Lat.	84,334	86,298	87,943	92,407	94,123	95,241
Ferret	Th.	0.74	0.68	0.64	0.64	0.62	0.59
	Lat.	89,552	91,382	93,531	91,373	92,938	94,172
Raytrace	Th.	0.69	0.66	0.62	0.61	0.58	0.56
	Lat.	86,354	88,210	89,213	88,393	89,378	90,127

Benchmarks		Graph size					
		Small			Large		
GAP	Parameters	PSO	TAPP	CoolMap	PSO	TAPP	CoolMap
Kron	Th.	0.67	0.63	0.61	0.60	0.58	0.56
	Lat.	83,292	85,721	86,519	92,142	94,216	95,312
Uniform	Th.	0.70	0.67	0.66	0.62	0.60	0.57
	Lat.	85,685	86,218	87,361	89,094	92,567	94,108

strategy shows better improvement for the NoCs having higher number of cores. A trade-off has also been established between the communication cost and the peak temperature, so that the designer can choose the potential solution.

Thermal heating problem is more severe in 3D NoC compared to 2D. In 3D systems, some layers are far away from heat sink. Thus, removing the hotspots from layers is a more challenging task. The next chapter discusses such issues and proposes a few solutions.

References

Bader, D. A., & Madduri, K. (2006). Designing multithreaded algorithms for breadth-first search and st-connectivity on the Cray MTA-2. In *International Conference on Parallel Processing (ICPP)* (pp 523–530). Piscataway: IEEE.

Beamer, S., Asanovic, K., & Patterson, D. (2015). *The GAP benchmark suite*. arXiv:1508.03619 [cs.DC].

Bienia, C., Kumar, S., Singh, J. P., & Li, K. (2008). The PARSEC benchmark suite: Characterization and architectural implications. In *International Conference on Parallel Architectures and Compilation Techniques (PACT)* (pp. 72–81).

Catania, V., Mineo, A., Monteleone, S., Palesi, M., & Patti, D. (2016). Cycle-accurate network on chip simulation with Noxim. *ACM Transactions on Modeling and Computer Simulation, 27*(1), 4:1–4:25.

Cplex (2013). www.ibm.com/software/in/integration/optimization/cplex

Feero, B. S., & Pande, P. P. (2009). Networks-on-Chip in a three dimensional environment: A performance evaluation. *IEEE Transactions on computers, 58*(1), 32–45.

Graph500. (2018). *Graph500 benchmark*. www.graph500.org

Kahng, A. B., Li, B., Peh, L. S., & Samadi, K. (2012). ORION 2.0: A power area simulator for interconnection networks. *IEEE Transactions on Very Large Scale Integration (VLSI) Systems, 20*, 191–196.

Li, S., Ahn, J. H., Strong, R. D., Brockman, J. B., Tullsen, D. M., & Jouppi, N. P. (2013). The McPAT framework for multicore and manycore architectures: Simultaneously modeling power, area, and timing. *ACM Transactions on Architecture and Code Optimization, 10*(1), 5:1–5:29.

Magnusson, P., Christensson, M., Eskilson, J., Forsgren, D., Hallberg, G., Hogberg, J., et al. (2002). Simics: A full system simulation platform. *Computer, 35*(2), 50–58.

Mahajan, R. (2002) Thermal management of CPUs: A perspective on trends, needs and opportunities. In *Keynote Presentation at the 8th International Workshop on THERMal INvestigations of ICs and Systems.*

Martin, M. M. K., Sorin, D. J., Beckmann, B. M., Marty, M. R., Xu, M., Alameldeen, A. R., et al. (2005). Multifacet's general execution-driven multiprocessor simulator (GEMS) toolset. *SIGARCH Computer Architecture News, 33*(4), 92–99.

Moazzen, M., Reza, A., & Reshadi, M. (2012). CoolMap: A thermal-aware mapping algorithm for application specific networks-on-chip. In *Proceeding of Euromicro Conference on Digital System Design (DSD)* (pp. 731–734). https://doi.org/10.1109/DSD.2012.35

Sahu, P. K., Shah, T., Manna, K., & Chattopadhyay, S. (2014b) Application mapping onto mesh based network-on-chip using discrete particle swarm optimization. *IEEE Transactions on Very Large Scale Integration (VLSI) Systems, 22*(2), 300–312.

Skadron, K., Stan, M. R., Huang, W., Velusamy, S., Sankaranarayanan, K., & Tarjan, D. (2003). Temperature-Aware microarchitecture. In *Proceeding of IEEE International Symposium on Computer Architecture (ISCA)* (pp. 1–12).

Zhu, D., Chen, L., Pinkston, T. M., & Pedram, M. (2015). TAPP: Temperature-aware application mapping for NoC-based many-core processors. In *Proceeding of Design, Automation Test in Europe (DATE)* (pp. 1241–1244).

Chapter 7
Thermal-aware Design Strategies for the 3D NoC-based Multi-Core Systems

To enhance the performance of 3D NoC-based systems, we have proposed design techniques for 3D NoC-based systems, in Chaps. 3–5. However, thermal issues have become critical roadblocks in achieving highly reliable 3D systems. In the last chapter, we have presented a solution to such problem for 2D NoC-based systems using thermal-aware application mapping strategy. The thermal problem becomes more severe in 3D NoC-based systems compared to 2D NoC, due to the increased power density and lower thermal conductivity of inter-tier dielectrics. Furthermore, in 3D systems, the heat sink is far away from some of the layers. Excessive high temperature can significantly degrade the interconnect and device reliability which may, in turn, cause functional and timing faults, reduce the mean time to failure and speed up the ageing process in 3D systems. Another concern introduced by technology scaling is the increased leakage power dissipation. Higher temperature results in increased leakage power dissipation, due to its exponential dependency on temperature. Increased leakage power leads to higher total power consumption, which in turn generates more heat and creates a vicious cycle. Thus, the excessively high temperature has to be controlled by an upper bound provided by a designer.

The overall problem can be stated as follows:

Given the properties of an application (in terms of its task graph), a limited number of TSVs with a distance among the TSVs and a peak temperature, an optimum association between tasks and cores as well as placement of the limited number of TSVs has to be determined such that the weighted communication cost (BW × hop-count) is minimized.

The inputs to the problem are as follows:

- A task graph G which represents the application.
- A topology graph T which corresponds to the 3D NoC.
- Power profile of each task when assigned to a core.

© Springer Nature Switzerland AG 2020
K. Manna, J. Mathew, *Design and Test Strategies for 2D/3D Integration for NoC-based Multicore Architectures*, https://doi.org/10.1007/978-3-030-31310-4_7

- Power consumption each router and link.
- A floorplan for the NoC, represented as F.
- A peak temperature of the system.

1 Proposed Techniques

In this work, it has been assumed that the application will be mapped onto a homogeneous-tile oriented partially connected 3D-mesh-based NoC. A core and its corresponding router together form a tile. The first two constraints in the above mentioned problem have been addressed in Chaps. 3–5. The third constraint, that is, thermal issue has been addressed in this chapter. The chapter proposes two solutions for the problem. First, a classical application mapping approach has been used to solve the thermal safety problem of 3D NoC. Here, the application mapping procedure distributes the tasks in such a way that the peak temperature of the entire system is reduced. High power consuming tasks are placed nearer to the heat sink, compared to the low power consuming ones, in the successive layers of the 3D system. A judicious mixing of tasks can even out the temperature with nominal degradation in system performance. The other potential alternative for this problem is the temperature-aware physical design with the placement of vertical through vias. The vertical through vias are effective thermal conductors, one effective heat removal approach in 3D IC. It can remove the heat from stacked silicon layers to the heat-sink that is often on top of the stack. Such vias are called *thermal vias*. In this design approach, thermal vias are placed into a given extra space in the die without disturbing the performance of the system. However, the congestion problem in routing gets increased due to the insertion of thermal vias. Proper distribution of these thermal vias can be utilized to reduce the same. The salient features of the approaches are as follows:

1. A PSO-based application mapping solution has been proposed to take care of the thermal constraint with a given tolerance on communication cost.
2. The thermal vias have been properly distributed in the given space on the die using a PSO-based strategy, to take care of peak temperature of the die.

2 Temperature Calculation

In Chap. 6, the temperature of tiles in a chip has been calculated by multiplying the thermal resistance of each tile in the chip with their corresponding consumed power values. Then, the peak temperature of the tile is calculated. In this work, the thermal resistance of each layer has been extracted from the tool HotSpot (Skadron et al. 2003), similar to the strategy discussed in Chap. 6. Thereafter, the temperature of the entire system has been calculated using a multi-grid method. Thus, the peak temperature is measured using expression (7.1).

$$T_{peak} = \max \left\{ \max \left\{ T_1^1, T_2^1, \ldots, T_q^1 \right\}, \ldots, \max \left\{ T_1^m, T_2^m, \ldots, T_q^m \right\} \right\} \quad (7.1)$$

where, T_i^j denotes the temperature of ith unit in jth layer of the 3D chip.

In order to model temperature effect of thermal vias in a 3D chip, this work used extended HotSpot tool (Meng et al. 2012). Such a tool allows users to specify the resistivity and capacitance for any unit in the chip. In this work, the thermal resistance of each layer has been extracted using the extended-HotSpot tool. Next, thermal resistance and capacitance values have been changed according to the requirement of thermal vias in the unit of the chip.

3 Thermal-Aware 3D Application Mapping

A judicious mix of tasks depending upon their consumed power may lead to a better thermal scenario. It can be achieved with a nominal sacrifice in the communication cost. The sacrifice is due to the placement of highly communicating tasks away to achieve better thermal behaviour. A tolerance limit is provided on the communication cost. In this approach, at first, all tasks of an application are mapped onto the system without considering the thermal constraint. Such cost is called the best communication cost of the application. Then, a certain percentage of sacrifice is provided to the best communication cost to achieve better thermal profile for the system. This section describes the corresponding PSO-based approach.

3.1 PSO Formulation

PSO-based evolutionary strategy has been developed in 1995 (Kennedy and Eberhart 1995). It is a population-based stochastic search and optimization technique, each particle represents a potential solution. Evolution of particles over a generation is guided by both the self and the swarm intelligence.

(a) **Particle formulation and fitness calculation**: The individual particles of the formulated PSO consists of two parts (shown in Fig. 7.1)—*task part* and *TSV part*. The task part corresponds to the mapping part. It is a permutation of task numbers. However, TSV part represents the vertical connection of the routers in the NoC. Routers are numbered in increasing order from the lowest to the highest tier. In each tier, router numbers are assigned in a row-wise manner, starting from the top-left up to the bottom-right corner. Such a configuration has been shown in Fig. 7.1a. The corresponding particle has been shown in Fig. 7.1b. Here, task 16 gets mapped to core1, task 13 to core 2, and so on.

In this work, it has been assumed that routers at same locations are of the same type in each tier. Such an assumption is justified as TSV geometry will not

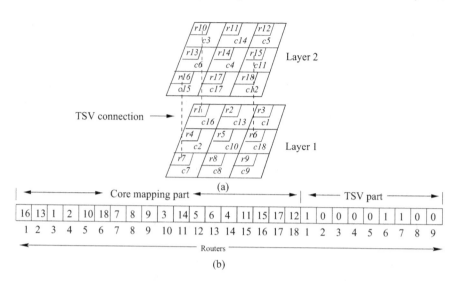

Fig. 7.1 Particle structure for thermal problem

allow two neighbouring routers in any layer to have TSV connections. Thus, it is better to put 3D routers at the same positions in each layer. The TSV part of the particle has been made as a bit-array of size equal to the number of routers at tier l. The bit being '1' indicates that the corresponding router in each layer has a vertical connection. Whereas, a '0' means that there is no vertical connection in the router. However, TSV constraints, like $x\%$ routers, in each layer can have TSVs and *t-hops* distance has to be maintained among the neighbouring routers with TSVs are maintained while placing the TSVs. Figure 7.1b shows the entire particle. During the implementation of PSO, the particle has been considered as a single dimensional array with appropriate care of the values contained in each cell.

In this approach a particle is measured by two quantities: communication cost (CC) and thermal cost. These two quantities jointly decide the quality of the particle. The communication cost is measured by expression (7.2), whereas the thermal cost is calculated using expression (7.1).

$$CC = \sum_i \sum_j \left(BW(c_i, c_j) \times Dist(u_i, u_j) \right) \tag{7.2}$$

where, $BW(c_i, c_j)$ denotes the bandwidth requirement between tasks c_i and c_j. $Dist(u_i, u_j)$ is the hop count in a shortest path between cores u_i and u_j, to which the tasks c_i and c_j have been mapped.

(b) **Local and global bests**: Each particle is associated with a local best which is a configuration with minimum communication cost and peak temperature among all configurations that particle has seen so far in the evolution process. In this proposed PSO-based approach, the communication cost of each local best is within the tolerance limit of the best communication cost. More precisely, we have given more preference in communication cost compared to temperature. That is, if two particles satisfy the given constraint, the particle with less communication cost will be chosen as the local best. The particle with the minimum peak temperature from the set of local best is declared as the global best.

(c) **Evolution of generation**: The initial population of particles is generated randomly and the corresponding fitness values are computed. Next, the global best particle is identified. Initially, the local best for each particle is set to the particle itself. The particles are evolve using a *swap* operation.

- *Swap operator*: In swap operation two indices, say i and j, of the particle p are taken as input and a new particle p_1 gets created. Both particles, p and p_1, are same excepting that the positions i and j of p are exchanged in p_1. Care has been taken to disallow swapping between core part and TSV part of a particle.

  Let p be a particle as shown in Fig. 7.2a. The swap operator $SO(3,5)$ exchanges the values at positions 3 and 5 in p to generate a new particle, as shown in Fig. 5.2b.

- *Swap sequence*: Set of swap operators create the swap sequence. For example, a swap sequence $SS = < SO(7, 1), SO(4, 3) >$ creates particle, P_{new}, by applying the operations on particle P in two steps. Figure 7.3a represents the particle P. Applying $SO(7,1)$ on P creates an intermediate particle, P_{mod}, shown in Fig. 7.3b. The swap $SO(3,4)$ on particle P_{mod} results in the new particle noted in Fig. 7.3c.

For the evolution of a particle, first, the swap sequences are identified to align it to its local best and the global best. The sequences are applied with some

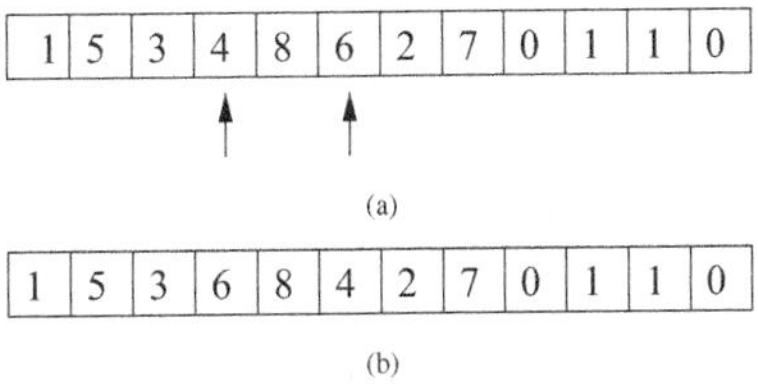

Fig. 7.2 An example of swap operation. (a) A particle before applying the *swap* operation (b) The particle after applying the *swap* operation

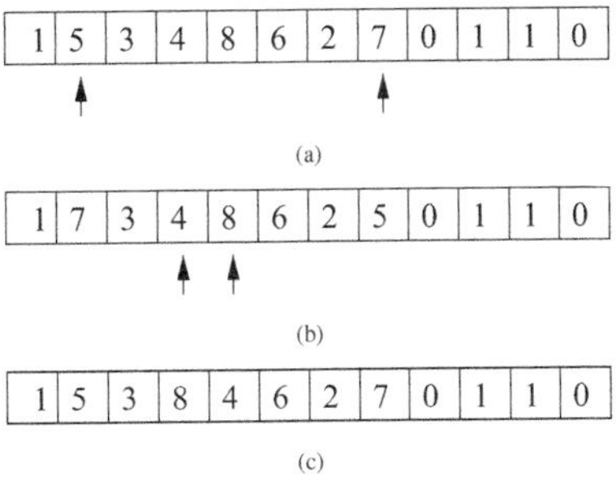

Fig. 7.3 An example of swap sequence operation. (a) A particle before applying the *swap* operation. (b) The particle after applying a *swap* operation. (c) The particle after applying the next *swap* operation

probabilities corresponding to the confidence factors. For our formulation, we have used the confidence factors to be 0.04 and 0.02, respectively, for local and global best alignment.

(d) **Augmentations to the basic PSO**: To achieve quality solution from the basic PSO strategy, the following augmentations have been integrated:

- *Usage of better random number generator*: The solution quality of PSO depends heavily on the property of used random number generator. In C-library routines for random number generation, rand() and rand48() use Linear Congruential Generator (LCG) (Saito and Matsumoto 2008). However, in this work, we have used the thread-safe single instruction multiple data-oriented fast Mersenne twister (SIMT) pseudorandom number generation technique (Saito and Matsumoto 2008). Such a technique provides several advantages over LCG. In particular, it has larger period (up to $2^{216091} - 1$), compared to ($2^{31} - 1$) for LCG (Saito and Matsumoto 2008), better equidistribution and quick recovery from 0-excess initial state. Compared to other statistically reasonable generators, it is faster and useful, when huge random values are required (Tian and Benkrid 2009). The SIMT has passed several statistical testing including the diehard test of Marsaglia and the load test of Hellekalek and Wegenkittl (Matsumoto and Nishimura 1998).

- *Inversion mutation*: The PSO works better compared to Genetic Algorithm (GA) (Guilan et al. 2008). However, in PSO, as compared to GA, the mutation operator is not incorporated which can bring sudden changes into a solution and thus possibly exploring a promising unexplored part of the search space. We perform a mutation operation when PSO is found to be not improving over a fixed predefined number of generations. Such mutation may take it out of probable local optima. Thus, we have introduced a *inversion mutation* (IM) operator. In this technique, first, a break-point is randomly generated for the core part of the particle. The portion from this break-point to end is inverted and joined at the end of the part before break-point. Next, the same is performed for the TSV part. Figure 7.4 shows the operation of such an inversion mutation.

Fig. 7.4 An example of inversion Mutation operation. (**a**) A particle before applying the *mutation* operation. (**b**) The particle after applying the *mutation* operation

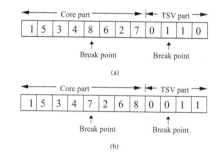

- *Initial population generation*: For an application with n tasks mapped onto a 3D-mesh having n cores distributed over m layers, the total number of possible mappings and TSV positions can be $n!$ and $(2^{n/m})$, respectively. Exploration of the potential region of this enormous search space depends to a great extent on the initial set of particles. Thus, in this work, we have used the deterministic initial population generation technique, described in Chap. 4, to help in the process. Such a strategy can generate a number of solutions equal to the number of cores in the NoC, quite fast. It works with randomly placed TSVs. The total number of TSVs has been restricted to 25% of total available positions. Also, TSVs are placed at least two hops away from each other. Next, for each core in the NoC, a solution in generated by starting the mapping process at this core node. The best one among them, along with the associated TSV positions contributes to the creation of one intelligent particle to be included in the initial population. The process has been repeated to create intelligent particles of 5% of the total population of the PSO. The rest of the particles are generated randomly.
- *Multiple PSOs*: In this augmentation, PSO has been run several times to improve the solution. Suppose that at the end of ith run of PSO, the local best for kth particle be $pbest_i^k$, and the global best be $gbest_i$. In the $(i + 1)$th pass of PSO, it starts with a new set of particles. However, the local and global best information are transferred from the ith to $(i + 1)$th PSO. The maximum number of the PSO runs to be executed has been set as follows:

  1. A user-defined value for the maximum number of PSO runs. In this experiment, it has been taken as 200 PSO runs.
  2. The global best fitness does not change in the last 20 PSO runs.

The proposed approach takes two inputs from the designer: a tolerance on the best communication cost and a bound on the peak temperature of the system. First, the best communication has been obtained without considering any temperature constraint. Fitness (communication cost) of each particle has been calculated. The local best is determined. The global best is set to the particle in the local best set having the minimum fitness. Thus, the best particle found in the last stage of PSO can be directly copied into the initial stage of the PSO with taking care of peak temperature. Furthermore, the local bests found in that PSO are also included in the initial stage of the current PSO which qualify the tolerance criteria on communication cost. The remaining particles in the proposed PSO are generated randomly.

4 Thermal-Aware 3D NoC Design Using Thermal-Vias

One of the biggest challenges of 3D NoC-based systems design is heat dissipation. The heat, from the core of a 3D NoC-based systems, flows through layers of low conductive dielectric, inserted between device layers, to reach the heat sink. To reduce the excessive heat in 3D NoC-based systems, thermal vias can be used to carry the heat from the layers, far away from the heatsink, to the nearer layer at the heatsink (Cong and Zhang 2005; Cong et al. 2004). However, such thermal vias consume valuable area in the die which in turn increases the routing cost. Judicious placement of such thermal vias can reduce the peak temperature and routing cost. A PSO-based strategy has been used to properly deploy such thermal vias. An area, on the chip, has been reserved in each tile of the NoC, for this purpose. More precisely, such an area is limited by the maximum value for each tile. Thus, proposed PSO takes two inputs, the peak temperature and the total area value used for thermal vias (in terms of percentage of total chip area without thermal vias) as the constraints.

4.1   PSO Formulation

The particle structure for the above problem has been shown below:

(a) **Particle formulation and fitness calculation**: The particle, corresponding to the problem, has been shown in Fig. 7.5 for $3 \times 3 \times 2$, 2-tire NoC. It consists of three parts: task part, TSV part and area part for thermal vias. The task part corresponds to the permutation of task numbers identifying the task mapped to a core, whereas, TSV part represents the location of vertical connection in the router. It has been assumed that routers at a similar position are of the same type in each layer and a minimum distance has also been considered between neighbouring routers with the vertical connection. The area part represents the percentage of the reserved area in the tile to be used for thermal vias. It has also been assumed that tiles at a similar position are of the same type, in each layer. The entries in the area part are in the range 0–1. Figure 7.5b shows a full particle structure. For the sake of implementation, a particle has been considered to be

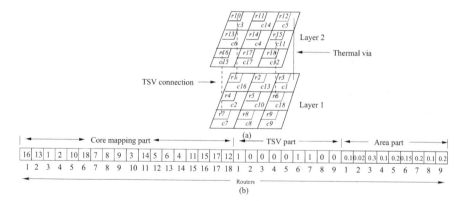

Fig. 7.5 Particle structure for thermal-aware 3D NoC-based system design using thermal vias. (**a**) A vertically-partially-connected 3D-mesh-NoC with thermal vias. (**b**) Particle structure for the configuration in (**a**)

a single array with appropriate care taken regarding the values contained in its cells. The location of the array indicates the router number. Further, a tile is also identified by the router number. Fitness of each particle is measured by three quantities: communication cost, peak temperature and the total area used for thermal vias. The communication cost of each particle is measured by the expression (7.2). The peak temperature of the 3D NoC has been measured by the expression (7.1).

In this work, copper-TSVs have been used as the thermal vias. As a result, we have calculated the percentage of area in the tile covered by TSVs ($A_{\%withTSVs}$) and the area without TSVs ($A_{\%w/oTSVs}$). The corresponding resistivity of the unit with thermal via has been calculated using the expression (7.3) (similar to Meng et al. (2012)). Such a resistivity has been updated before calculating the temperature of the system. The area cost is calculated by summing up the area used by the thermal vias. The cost must be honoured the tolerance given on the total area, without thermal vias.

$$R_{joint} = \frac{1}{\frac{A_{\%w/oTSVs}}{R_{Layer}} + \frac{A_{\%withTSVs}}{R_{CU}}} \tag{7.3}$$

where, R_{Layer} denotes the resistivity of the layer (thermal interface material (TIM) or Silicon) and R_{CU} represents the resistivity of TSVs (Copper = 0.0025 mK/W). The resistivity of TIM and Silicon are 0.25 mK/W and 0.01 mK/W, respectively.

(b) **Local and global bests**: In this PSO-based technique, each particle is also associated with a configuration (local best) which has the minimum peak temperature and area requirement for the thermal vias that particle has seen so far in the evolution process. In this work, we have given more preference to temperature, compared to area. That is, if two particles satisfy the given constraint, the particle with less temperature will be selected. Such configuration is known as local best (*lbest*) of the corresponding particle. The configuration with minimum peak temperature and area requirement for thermal vias in the local best set is declared as global best (*gbest*) of the current generation.

(c) **Evolution of generations**: The area part of the particle evolves through the strategy followed in continuous PSO domain. The particle is attracted by its local as well as global bests particle, in the generation and the particle updates its velocity and position using the following expressions:

$$v_{k+1,i} = c_1 \times v_{k,i} + c_2 \times r_1 \times (lbest_i - x_{k,i}) + c_3 \times r_2 \times (gbest_k - x_{k,i})$$

$$\tag{7.4}$$

$$x_{k+1,i} = x_{k,i} + v_{k+1,i} \tag{7.5}$$

where, $v_{k+1,i}$ denotes the velocity of the particle i at $(k+1)$th iteration, $x_{k,i}$ is the current particle solution, the random numbers r_1 and r_2 are in the range of 0 to 1. The self-confidence (cognitive) and swarm confidence (social) factors have been represented by c_2 and c_3, respectively. The inertia is represented by c_1. For controlling the velocity of a particle, a maximum velocity v_{max} is imposed on particle's velocity. So if any particle moves with a velocity that exceeds the v_{max} then that particles velocity is curtailed to v_{max}. However, task mapping and TSV placement do not evolve as they are fixed. Thus, performance of the 3D NoC-based system do not degrade as in previous solution.

The convergence condition of this PSO can be expressed as (Guilan et al. 2008)

$$(1 - \sqrt{c_1})^2 \leq c_2 + c_3 \leq (1 + \sqrt{c_1})^2 \tag{7.6}$$

Thus, we have worked with various values of c_1, c_2 and c_3. The experiment result in this chapter based on values $c_1 = 1$, $c_2 = 0.5$ and $c_3 = 0.5$.

(d) **Augmentations to the basic PSO**: To improve the solution quality, given by the basic PSO, several augmentation have been used, such as, *efficient random number generation*, *inversion mutation operation* and *multiple PSO*. All such techniques have been discussed in Sect. 3.1. The maximum number of the PSO runs to be executed has been set as follows:

1. A user-defined value for the maximum number of PSO runs. In this experiment, it has been taken as 200.
2. The global best fitness meets the thermal constraint.

5　Experimental Results

This section presents the experimental results of different NoC benchmarks. The simulation environment used in this work is similar as follows. A fixed ambient temperature of 45 °C has been assumed (Mahajan 2002) in the current work. The power consumption for NoC components like routers and links have been calculated using *Orion* (Kahng et al. 2012), whereas, core power values have been chosen similar to (Tsai and Kang 2000) (10–60 W/cm^2). The power consumption of NoC components and core combinedly form the tile power. The applications are mapped onto 3D NoC with two and four layers. The results are organized as follows: First, results for thermal-aware 3D NoC design with application mapping strategy has been shown with the varying tolerance limit on communication cost. Next, results for thermal-aware NoC-based systems design with thermal vias have been presented with varying area requirement.

5.1 Thermal-Aware 3D NoC-Based Systems Design with the Application Mapping Strategy

In this section, Table 7.1 shows the communication cost and peak temperature comparison by varying tolerance limit on communication cost and given the peak temperature as a constraint, for 3D NoC-based systems with two and four layers. From Table 7.1, it is observed that temperature reduces as the communication cost increases. In this experiment, we have considered the tolerance limit on communication cost as 5%, 10% and 15%. The cell marked as a, b indicates the communication cost and peak temperature of a solution. The temperature and communication cost always honouring the given constraints. The constraints are marked as (m, n) in the table, where m indicates tolerance on communication cost and n indicates a peak temperature constraint. Therefore, the designer can choose the corresponding solution based on their requirements.

5.2 Thermal-Aware 3D NoC-Based Systems Design Using Thermal Vias

In this experiment, the peak temperature of the systems and tolerance limit on chip area, for thermal vias, have been taken as input. The communication cost and peak temperature have been presented in Table 7.2 honouring the peak temperature limit

Table 7.1 Comparison of communication cost with different tolerance limit on communication cost and a given peak temperature

Layers	NoCs	Tolerance on comm. cost and Peak temperature		
		(5%, 85 °C)	(10%, 80 °C)	(15%, 75 °C)
Two	PIP	800, 83	825, 77	910, 73
	263ENC-MP3DEC	235.23, 81	251.12, 75	265.31, 72
	VOPD	4218, 80	4375, 76.11	4831, 72.23
	DVOPD	9919, 81	10,347, 78.15	11,014, 73.3
	G17	36,125.13, 83.2	37,461.28, 77.2	39,102.53, 74
	G18	6150.17, 82	6617.42, 76	7185.9, 73.1
	G25	102,516.47, 81	107,345.18, 77	110,853.75, 72
	G27	48,176.16, 81.32	49,827.52, 78.3	53,181.67, 74.1
Four	PIP	901, 84.31	929, 79	1012, 74
	263ENC-MP3DEC	235.23, 84.3	251.12, 78.23	265.31, 73.31
	VOPD	4376, 83.23	4512, 79.12	4817, 74
	DVOPD	9978, 82	10,236, 79	10,521, 74.12
	G17	36,321, 84.35	38,167, 79.56	40,762, 74
	G18	6415.76, 83	6734.94, 79.1	7123.14, 74.3
	G25	101,457.23, 84.3	107,341.13, 78	114,321.43, 74.2
	G27	45,129.74, 83	50,123.31, 79.6	53,429.26, 74.35

Table 7.2 Comparison of communication cost and peak temperature for thermal vias with different percentage of reserved area

Layers	NoCs	Tolerance on area and peak temperature		
		(10%, 130 °C)	(15%, 125 °C)	(20%, 120 °C)
Two	PIP	768, 81.21	768, 76.15	768, 73.37
	263ENC-MP3DEC	230.43, 78.14	230.43, 120.67	230.43, 72.21
	VOPD	4119, 82.35	4119, 78.17	4119, 72.21
	DVOPD	9554, 83.39	9554, 77.26	9554, 73.15
	G17	35,375.93, 82.29	35,375.93, 78.38	35,375.93, 73.23
	G18	6094.11, 83.32	6094.11, 77.65	6094.11, 72.67
	G25	99,815.93, 83.11	99,815.93, 78.78	99,815.93, 73.38
	G27	47,121.62, 83.76	47,121.62, 78.32	47,121.62, 72.12
Four	PIP	896, 82.12	896, 78.11	896, 74.26
	263ENC-MP3DEC	230.45, 78.32	230.45, 120.33	230.45, 73.38
	VOPD	4237, 84.68	4237, 79.52	4237, 74
	DVOPD	9768, 84.67	9768, 79.35	9768, 74.12
	G17	36,565, 84.18	36,565, 79.21	36,565, 74.38
	G18	6222.23, 84.32	6222.23, 78.66	6222.23, 74
	G25	99,126.77, 84.27	99,126.77, 79.11	99,126.77, 74.67
	G27	46,380.91, 84.61	46,380.91, 79.56	46,380.91, 74.35

and tolerance on the area. From the table, it is seen that the temperature reduces with the increase in tolerance limit on the area. For 3D NoC-based systems, with 2 layers, the proposed approach provides communication cost and peak temperature as 4119 and 124.35 °C, respectively, while 10% tolerance has been considered on the total area for thermal vias and peak temperature constraint as 130 °C for the application VOPD. Moreover, this approach reduces the temperature better compared to our proposed previous mapping-based strategy. Furthermore, the reduction in peak temperature of the system using thermal vias is not unlimited. It is loosely bounded by the minimum temperature of the design before inserting the TSVs.

6 Conclusion

This chapter presents two techniques to design of 3D NoC-based systems with ensuring thermal safety, using PSO with several augmentations. The first approach is based on an application mapping strategy to reduce the peak temperature of the system by giving some nominal sacrifice in communication cost. The second one is based on inserting thermal vias into the systems in the provided reserved space on the chip without sacrifice in communication cost. The experimental results show that the second approach reduces the temperature well compared to the first one. However, it is up to the designer to choose a suitable solution based on the requirement.

Thermal heating, during test of such a system, can reduce the yield and also increase the testtime. To deal with those problems, some of the strategies as proposed are explained in the next chapter.

References

Cong, J., Wei, J., & Zhang, Y. (2004). A thermal-driven floorplanning algorithm for 3D ICs. In *Proceedings of the 2004 IEEE/ACM International Conference on Computer-Aided Design* (pp. 306–313).

Cong, J., & Zhang, Y. (2005). Thermal via planning for 3-D ICs. In *Proceedings of the 2005 IEEE/ACM International Conference on Computer-Aided Design* (pp. 745–752).

Guilan, L., Hai, Z., & Chunhe, S. (2008). Convergence analysis of a dynamic discrete PSO algorithm. In *2008 First International Conference on Intelligent Networks and Intelligent Systems* (pp. 89–92).

Kahng. A. B., Li, B., Peh, L. S., & Samadi, K. (2012). ORION 2.0: A power area simulator for interconnection networks. *IEEE TransVery Large Scale Integration (VLSI) Systems, 20,* 191–196.

Kennedy, J., & Eberhart, R. (1995). Particle swarm optimization. In *Proceedings of ICNN'95-International Conference on Neural Networks* (pp. 1942–1948)

Mahajan, R. (2002). Thermal management of CPUs: A perspective on trends, needs and opportunities. In *Keynote Presentation at the 8th International Workshop on THERMal INvestigations of ICs and Systems*

Matsumoto, M, & Nishimura, T. (1998). Mersenne twister: A 623-dimensionally equidistributed uniform pseudo-random number generator. *ACM Transactions on Modeling and Computer Simulation, 8*(1), 3–30.

Meng, J., Kawakami, K., & Coskun, A. (2012). Optimizing energy efficiency of 3-D multicore systems with stacked DRAM under power and thermal constraints. In *Proceedings of the 49th Annual Design Automation Conference* (pp. 648–655).

Saito, M., & Matsumoto, M. (2008). SIMD-oriented fast Mersenne twister: A 128-bit pseudorandom number generator. In A. Keller, S. Heinrich, & H. Niederreiter (Eds.), *Monte Carlo and Quasi-Monte Carlo methods*. Berlin: Springer.

Skadron, K., Stan, M. R., Huang, W., Velusamy, S., Sankaranarayanan, K., & Tarjan, D. (2003). Temperature-aware microarchitecture. In *Proceedings 30th Annual International Symposium on Computer Architecture, 2003* (pp. 1–12).

Tian, X., & Benkrid, K. (2009). Mersenne twister random number generation on FPGA, CPU and GPU. In 2009 NASA/ESA Conference on Adaptive Hardware and Systems(pp. 460–464).

Tsai, C., Kang, S. (2000) Cell-level placement for improving substrate thermal distribution. *IEEE Transactions on Computer-Aided Design of Integrated Circuits and Systems, 19,* 253–266.

Chapter 8
Thermal-Aware Test Strategies for NoC-Based Multi-Core Systems

We have seen several techniques to design the NoC-based multi-core systems in the previous chapters. To get confidence of the current operation of such system, it is required to test the manufactured chip. The task of manufacturing test of multi-core systems for its complete functionality is complex and time consuming Kiamehr et al. (2018). Thus, the high stress to reduce time-to-market have made the test engineers to focus primarily onto the reduction of testtime, including thermal safety. A good test technique can improve the yield and reduce the testtime.

In SoC-based system, designers integrate different cores taken from different vendors. Vendors provide not only cores, but also set the test patterns to test. Furthermore, some core can support preemption whereas others do not during the test process. A core requiring that the entire test be performed in a continuous fashion is called a non-preemptive core, while the other one allows preemption during test. Such test patterns can ensure the correctness of core, not the integrated system. The test set provided by the vendor needs to applied to the input of the core and responses are to be observed. This challenge as the input–output lines of the cores, integrated inside the chip, are not directly accessible from the system input–output pins. In SoC environment, such problem can be resolved by providing an extra dedicated test access mechanism (TAM) for the chip. TAM is accessible from system input–output pins of the system. The cores are accessible through the TAM. Several test access mechanisms have been proposed. However, one (or more) dedicated bus(es) is the most suitable one. In test mode, test patterns are carried by the TAM and applied to the core, instead of functional inputs. The generated test responses are carried accordingly to the output port of the system. To ensure the correctness of the entire system, core level testing has to be done exhaustively. This leads to huge testtime for moderate to complex SoCs. Hence, testtime reduction is a very important issue. Test engineers often perform parallel test to reduce the testtime. The overall TAM structure is divided into a number of buses to carry out the test in parallel.

© Springer Nature Switzerland AG 2020
K. Manna, J. Mathew, *Design and Test Strategies for 2D/3D Integration for NoC-based Multicore Architectures*, https://doi.org/10.1007/978-3-030-31310-4_8

However, in a NoC environment, usage of extra TAM is not advisable. The NoC-based system contains a network for communication. Such network can be reused to carry the test information for the cores. In NoC-based system, some dedicated cores are used as input–output (I/O) ports/pairs for testing the entire system. The Automatic Test Equipment (ATE) connects to the system by these ports to supply test patterns and collets the responses.

1   Testtime and Temperature Calculation

In this section, we present the technique to compute the testtime and chip temperature for a multi-core system. It has been assumed that the cores have been wrapped in test-wrappers using Iyengar et al. (2002). In this work, wrapper has been designed for 32 *bits* NoC channel. Figure 8.1 represents a 3D multi-core systems, where each 2D-mesh-layer is vertically connected, using the Through Silicon Via (TSV)—most potential solution for vertical connection Dubois et al. (2013). Due to area overhead for TSVs, all routers, in 2D mesh cannot have vertical connections. Each pair of adjacent 2D mesh layers are partially connected using the TSVs. In

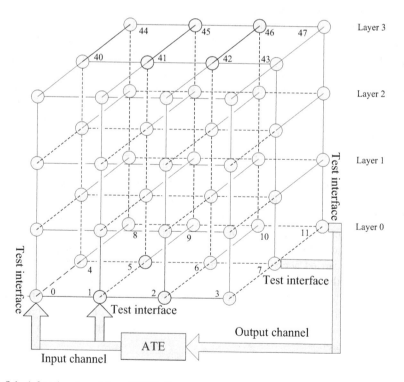

Fig. 8.1 A $3 \times 4 \times 4$ mesh based 3D NoC system

Liu et al. (2011a), the authors have minimized the vertical connection by sharing a TSV among the four neighbouring cores. The overall connection structure is called symmetric partially connected 3D mesh Liu et al. (2011a). We have considered this connection structure in our work. In Fig. 8.1, two IO pairs are denoted by 'S1' and 'K1', and 'S2' and 'K2'. The channel of Automatic Test Equipment (ATE) are connected via IO pairs. Let us say core 42 is tested by the IO pair ('S2', 'K2'), core 1 (source) will feed the test pattern to core 42. The test packet will move through the NoC following a deadlock free *Elevator-first*-routing technique for partially 3D NoC Dubois et al. (2013) (*xy*-routing is used for 2D NoC). In elevator-first routing algorithm, router 0 is the nearest one, having the elevator. So, test packet will be sent using router 0, then router 12, 24 and 36. Now, the test packet will be sent to router 42 using *xy*-routing algorithm. Test responses will be collected at core connected with router 11 (sink) using *elevator-first*-routing algorithm. The elevator routing algorithm chooses the routers 43, then 47, 35 and 23 for sending the response to sink 11. Test and response packets move in a pipelined fashion through this path.

In preemptive test scheduling, test packets are distributed as several chunks (discussed in Sect. 3.1). Testtime for a chunk can be calculated as follows. Let p denote the number of test patterns in the chunk. Maximum scan-chain length (number of flits per packet) of the core be denoted as l and the distance (in terms of hops) from source to core and core to sink be denoted as $h_{s \to c}$ and $h_{c \to k}$, respectively. As each flit is unpacked in one clock cycle and flits move in pipelined fashion, the time to send a test packet from source to the core plus the time to collect response in the sink is given by: $h_{s \to c} + (l - 1) + 1 + h_{c \to k} + (l - 1)$.

During testing, consecutive flits are applied to the core. So, the minimum inter-arrival time between consecutive flits is given by *Max{time required to shift-in a flit into the core, time required to shift-out a response from the core} + unit cycle to unpack the flit + time taken to generate the response*. Therefore, entire testtime ($\mathbb{T}$) for a particular chunk is given by

$$\mathbb{T} = h_{s \to c} + (l - 1) + 1 + \left[1 + Max\{h_{s \to c} + (l - 1), h_{c \to k} + (l - 1)\} \right]$$
$$\times (p - 1) + h_{c \to k} + (l - 1)$$
$$= \left[1 + Max\{h_{s \to c}, h_{c \to k}\} + (l - 1) \right] \times p + \left[Min\{h_{s \to c}, h_{c \to k}\} + (l - 1) \right]$$
$$(8.1)$$

Temperature of a tile/IP-block depends on its power consumption and its position in the floorplan. Considering such configuration, *HotSpot* Huang et al. (2006), an efficient thermal modeling tool has been developed to measure the thermal effect at IP-block level. A simple compact model calculates the temperature of each IP-block by considering the heat dissipation within the block and also the effect of heat transfer among the IP-blocks, based on the RC model. The thermal resistance, $R_{i,j}^{th}$ of the IP-block IP_i with respect to IP_j can be defined as: $R_{i,j}^{th} = \Delta T_{i,j}/\Delta P_j$, where $\Delta T_{i,j}$ represents the increment in temperature of IP_i due to ΔP_j power dissipation at IP_j. Total number node in the floorplan is denoted by q. Therefore, the thermal resistance matrix can be defined as: matrix $R^{th} = [R_{i,j}^{th}]_{q \times q}$.

For a given set of power values of individual node in floorplan, the temperature of each tile can be computed as: $[T_i]_{q \times 1} = [R_{i,j}^{th}]_{q \times q} \times [P_i]_{q \times 1}$, where P_i and T_i denote the power and temperature values of the ith IP-block, IP_i. The peak temperature of the die can be calculated as

$$T_{peak} = \max\{[T_i]_{q \times 1}\} \tag{8.2}$$

For specific floorplans, the values for matrix, Rth, can be computed through HotSpot Huang et al. (2006), and this is the same matrix used internally by HotSpot. Similarly, in 3D multi-core systems, thermal resistance matrix for each layer has been computed as similar in *HotSpot*, for specifics 3D floorplan. The corresponding temperature of each block has been calculated and the peak temperature has been obtained. The detailed setup for HotSpot configuration, especially for 3D system, is found in Sect. 6.

2 Problem Formulation of Preemptive Test Scheduling

A core supporting preemption can be scheduled in either preemptive or non-preemptive manner. The preemptive scheduling approaches often result in lower testtime than the non-preemptive one, because tests can be preempted to avoid resource conflicts that can create idle time durations in test resources. In preemptive test scheduling policy, this idle time can be used to test other cores. The preemptive test strategy does not significantly increase the ATE's control complexity. It stores the test patterns in the same way as in non-preemptive strategy, except that the sub-sequences of test patterns are reordered in ATE memory. Therefore, a judicious mix of sequences testing preemptive and non-preemptive cores can have better flexibility in utilizing testtimes which may otherwise remain idle, due to resource conflicts. Furthermore, to reduce the testtime, the test engineers often perform parallel core test by using the property provided by the on-chip network. However, this increases the power consumption during the test. Power consumption during test is significantly higher than the normal mode of operation. Thus, to reduce the power consumption during test, it is often required to perform power-aware test scheduling. In such a technique, the test scheduler always honour the power constraint by adjusting the supported clock rate of cores during test. However, it does not consider the thermal heating problem of the system while testing. Thermal heating can create delay variations in system which may cause a bad system to pass or a good system to fail the test process leading to yield loss. Thermal heating reduces system reliability which also leads to failure of the system. The leakage power consumption increases due to the thermal problem which may damage the system.

The input to the test scheduling problem consists of the following:

- The set $C = \{C_i : 1 \le i \le N\}$ of N number of cores in the system.

- Each core C_i is tested by a set of test patterns, T_i. The power consumption during the test of core C_i is PWR_i. The test can be non-preemptive or preemptive in nature.
- The maximum limit on system-level power consumption for the entire test session is P.
- A set $F = \{F_i : f/m \leq i \leq m \times f\}$ of $(2m - 1)$ of frequencies that can be generated from tester clock with frequency f, where m is the rate for faster and slower clocks.
- A set $IO = \{IO_i : 1 \leq i \leq k\}$ of k cores marked as input–output pairs.
- The chip floorplan G.
- Routing algorithm.

The integrated test scheduling problem can be stated as follows:

Determine the distribution of clock frequencies and IO pairs, optimum number of preemptions and test start time for each core, such that (1) testtime and resource conflicts get minimized (2) the schedule always satisfies the system-level power constraint and provides a trade-off between testtime and peak temperature during test.

To solve such problem, power-aware testing, alone, cannot guarantee system safety while testing. Thus, this chapter presents testing strategies for NoC-based system, including 3D environment, to improve yield loss and testtime while considering thermal safety with honouring power constraint. In this work, we have considered the maximum and minimum allowable clock rates $2f$ and $f/2$, respectively, that is, $m = 2$. Therefore, available frequency list is $F = \{f/2, f, 2f\}$. The salient feature of the approach are as follows.

3   ILP Formulation for Preemptive Test Scheduling

In this section, we present an ILP formulation to minimize the testtime for preemptive testing of cores. Test packets of each core are arranged into different pools. The detailed structure about the pool is discussed in below. The parameters and variables used in the ILP formulation have been noted in Table 8.1. To reduce the complexity, the ILP formulation does not take into consideration of the power and thermal issues. Power and thermal constraints have been incorporated in the PSO-based strategy, noted in Sect. 4.

3.1   Pools Structure

The complete set of test packets of a core are arranged into a particular pool. For a core, there are several pools and each pool is uniquely identified by a *pool index* p. In each pool, the complete set of test packets are ordered as test chunks, and

Table 8.1 Variables used for ILP formulation

Variables	Definitions
N_c	Total available cores
N_I	Total available IO pairs
N_f	Total available frequencies
N_t^c	Total number of test packets of core c
$T_{c,p}^{I,k}$	Testtime of core c for a chunk in the pool p, while IO pair I and kth frequency have been assigned for the test, which is precalculated
$BIGNUM$	Testtime when all cores are tested sequentially
F_c^k	=1, if the kth frequency is assigned to test the core c
	=0, Otherwise
$b_{c,p,a}^I$	=1, if a chunks are taken from pool p for the test of core c and are transported through
	an IO pair, I
	=0, Otherwise
$S_{c,p,a,x}^I$	The start time for the test of core c with the x instance of a chunks which are taken from pool p
	and such instance of chunk is transported through an IO pair I
$Z_{c_1,I_1}^{c_2,I_2}$	=1, if cores c_1 and c_2 are tested through IO pairs I_1 and I_2 are conflict
	=0, Otherwise
	Which has been precalculated
H_c^I	=1, if core c is assigned with an IO pair I
	=0, Otherwise

each chunk has a length, which is same as the pool index. That is, each chunk of pool p contains p test packets. The total available pools of each core is same as its number of test packets. For example, suppose, core c has four test packets. There will be four pool of test packets for core c. Each chunk in the 1st pool is of length unity; each chunk contains single test packet and the pool contains all four test packets. Similarly, 2nd pool contains all four test packets which are arranged into the chunks of length two. That is, 2nd pool contains two test chunks and each test chunk holds two unique test packets. Similarly, other pools are arranged. The last pool, with index four, has a chunk of length four. The pool of packets for core c has been shown in Fig. 8.2. If the entire test of core c has been done with four packets taken from the first and second pools, two chunks will be taken from the pool 1 and one chunk from pool 2. At most 'a' chunks can be taken from a pool p, such that $(a \times p) \leq N_t^c$. However, all test chunks, for a core (c), are tested by an unique I/O pair (H_c^I), selected from the N_I available I/O pair and a unique test clock frequency (F_c^k) selected from N_f available frequencies, selected by the test scheduler. The test scheduler, also, assigned the start time ($S_{c,p,a,x}^I$) of each chunk and calculates testtime ($T_{c,p}^{I,k}$) of each chunk, with assigned I/O pair and test

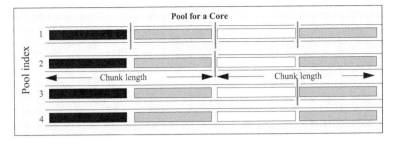

Fig. 8.2 The core with four test packets and its pool structure

frequency, using Eq. (8.1). The test scheduler gets total testtime for a core with the test chunk having maximum testtime. Similarly, test scheduler calculates testtime for N_c cores. The core with maximum testtime provides the entire testtime for the entire systems. Now, the objective is to minimize this overall testtime, which is elaborated in the following:

3.2 Objective Function

To minimize the overall testtime (TT) of a NoC-based multi-core system, the objective function can be defined as follows:

$$Min\ TT : Max\left(S^I_{c,p,a,x} + \sum_{k}^{N_f}\left(F^k_c \times T^{I,k}_{c,p}\right)\right), \forall core\ c,\ p,\ a,\ x \qquad (8.3)$$

To represent the above objective in ILP form, we have introduced a constraint shown in expression (8.4).

$$TT \geq \sum_{I}^{N_I} b^I_{c,p,a} \times \left(S^I_{c,p,a,x} + \sum_{k}^{N_f}\left(F^k_c \times T^{I,k}_{c,p}\right)\right) \qquad (8.4)$$

$$O^{I,k}_{c,p,a} = b^I_{c,p,a} \times F^k_c \ ; \ O^{I,k}_{c,p,a} \leq b^I_{c,p,a} \ ; \ O^{I,k}_{c,p,a} \leq F^k_c \ ; \ O^{I,k}_{c,p,a} \geq F^k_c - \left(1 - b^I_{c,p,a}\right) \qquad (8.5)$$

$$Q^{I,k}_{c,p,a,x} = b^I_{c,p,a} \times S^I_{c,p,a,x} \ ; \ Q^{I,k}_{c,p,a,x} \leq BIGNUM \times b^I_{c,p,a} \ ; \ Q^{I,k}_{c,p,a,x} \leq S^I_{c,p,a,x} \qquad (8.6)$$

$$Q^{I,k}_{c,p,a,x} \geq S^I_{c,p,a,x} - \left(1 - b^I_{c,p,a}\right) \times BIGNUM \ ; \ Q^{I,k}_{c,p,a,x} \geq 0 \qquad (8.7)$$

To linearize the expression (8.4), we have introduced one binary variable, $O_{c,p,a}^{I,k}$ to replace the term $b_{c,p,a}^{I} \times F_c^{k}$ and added four constraints, (8.5). To replace the other term $b_{c,p,a}^{I} \times S_{c,p,a,x}^{I}$, we have introduced another variable, $Q_{c,p,a,x}^{I}$ and added four constraints, (8.6)–(8.7).

3.3 Constraints

To meet the above objective, the following constraints have been introduced which must be satisfied during the optimization process.

IO Pair and Test Clock Assignment Constraint Each core can be tested by exactly one IO pair out of N_I IO pairs, which has been enforced by expression (8.8).

$$\forall c, \quad \sum_{I}^{N_I} H_c^{I} = 1 \tag{8.8}$$

$$\forall c, \quad \sum_{k}^{N_f} F_c^{k} = 1 \tag{8.9}$$

where, I denotes the available IO pairs. Each core has been assigned to one clock rate out of N_f rates. To maintain such constraint, the binary variable F_c^{k} and expression (8.9) have been used. where, k denotes the number of frequencies core c can support.

Test Chunks Selection Constraints Let us say, to test a core c by using 'm' chunks from pool 'p' such that $(m \times p \leq N_t^c)$ and 'n' chunks from pool 'q' such that $(n \times q \leq N_t^c)$. However, for core c, there can be other pools, except 'p' and 'q', which are not considered in this test scenario. The selected test chunks will be send by using an IO pair I. The binary variable $b_{c,p,a}^{I}$ has been introduced for indicating whether IO pair I is taken or not for the test, by the selected 'a' chunks from pool 'p'. If 'a' chunks are taken from pool 'p' and IO I is used to send these test packets then $b_{c,p,a}^{I} = 1$. Otherwise $b_{c,p,a}^{I} = 0$. Such constraint is preserved by the expression (8.10). Similar kind of binary variables are also there for other pools. All chosen chunks from different pools must form the entire set of test packets for the core c. Such constraint is enforced by expression (8.11).

$$\forall p, c, \quad \sum_{I}^{N_I} \sum_{a}^{(a \times p) \leq N_t^c} b_{c,p,a}^{I} \leq 1 \tag{8.10}$$

$$\forall c, \quad \sum_{p=1}^{p \leq N_t^c} \sum_{I}^{N_I} \sum_{a}^{(a \times p) \leq N_t^c} (a \times p) \times b_{c,p,a}^{I} = N_t^c \tag{8.11}$$

Moreover, all chunks of core c are transported by an IO pair, I. This has been introduced by the expressions (8.8) and (8.12).

$$\forall c, \quad \sum_{I}^{N_I} H_c^I \times \left(\sum_{p=1}^{p \le N_t^c} \sum_{a}^{(a \times p) \le N_t^c} (a \times p) \times b_{c,p,a}^I \right) = N_t^c \qquad (8.12)$$

The constraint (8.12) is not linear. To linearize it, a binary variable $m_{c,i,j}^I$ has been introduced to replace the term $H_c^I \times b_{c,p,a}^I$. It takes a value of 1, if both H_c^I and $b_{c,p,a}^I$ are 1. Such condition can be enforced by adding four additional constraints, similar to (8.5).

Constraints for Sending the Chunks Simultaneously and Sequentially Simultaneous testing of cores may lead to resource conflicts due to usage of already assigned of IO pairs, router ports and links. Furthermore, test chunks a core must share the same IO pair; though they may be belong to different pools. Thus, those test chunks cannot be scheduled simultaneously. However, the test chunks of different cores can be scheduled simultaneously. The following constraints are enforced to meet those conditions:

Parallel Core Test Constraints During testing of cores, test chunks are transported through the assigned IO pair. More than one core can be tested simultaneously if they do not share any common resources, such as, IO pair, router ports and links of NoC. Testtime of two cores cannot be overlapped if they share any resource. Suppose, cores c_1 and c_2 are assigned to IO pairs I_1 and I_2, respectively, for the test. Assumed that the x_1-th and x_2-th are the instances of test packet of a_1 and a_2 chunks and which are sent simultaneously through the IO pairs I_1 and I_2. Furthermore, a_1 and a_2 chunks belong to pools p_1 and p_2 of cores c_1 and c_2, respectively. The testtime of these two cores do not overlap if only if either one of the following conditions is valid.

$$S_{c_1,p_1,a_1,x_1}^{I_1} \ge S_{c_2,p_2,a_2,x_2}^{I_2} + \sum_{k}^{N_f} \left(F_{c_2}^k \times T_{c_2,p_2}^{I_2,k} \right) \text{ (or)} \qquad (8.13)$$

$$S_{c_2,p_2,a_2,x_2}^{I_2} \ge S_{c_1,p_1,a_1,x_1}^{I_1} + \sum_{k}^{N_f} \left(F_{c_1}^k \times T_{c_1,p_1}^{I_1,k} \right) \qquad (8.14)$$

A resource conflict is defined by the binary variable $Z_{c_1,I_1}^{c_2,I_2}$, when cores c_1 and c_2 are assigned to IO pairs I_1 and I_2 and has common resource to share. Now, to prevent the conflicting cores being scheduled at the same time can be formulated by using variable $Z_{c_1,I_1}^{c_2,I_2}$ as in expressions (8.15) and (8.16).

$$Z_{c_1,I_1}^{c_2,I_2}\left(S_{c_1,p_1,a_1,x_1}^{I_1} - S_{c_2,p_2,a_2,x_2}^{I_2} - \sum_{k}^{N_f}\left(F_{c_2}^{k} \times T_{c_2,p_2}^{I_2,k}\right)\right) \geq 0 \text{ (or)} \qquad (8.15)$$

$$Z_{c_1,I_1}^{c_2,I_2}\left(S_{c_2,p_2,a_2,x_2}^{I_2} - S_{c_1,p_1,a_1,x_1}^{I_1} - \sum_{k}^{N_f}\left(F_{c_1}^{k} \times T_{c_1,p_1}^{I_1,k}\right)\right) \geq 0 \qquad (8.16)$$

Moreover, the value for the variable, $Z_{c_1,I_1}^{c_2,I_2}$, has been precalculated by considering all possible conflict cases and provided to the ILP solver. Therefore, maintaining such constraint in ILP, unique binary variables m and n have been introduced, such that, $m + n = 1$. Then, the above expressions (8.15) and (8.16) can be written as

$$m\left(S_{c_1,p_1,a_1,x_1}^{I_1} - S_{c_2,p_2,a_2,x_2}^{I_2} - \sum_{k}^{N_f}\left(F_{c_2}^{k} \times T_{c_2,p_2}^{I_2,k}\right)\right) +$$
$$n\left(S_{c_2,p_2,a_2,x_2}^{I_2} - S_{c_1,p_1,a_1,x_1}^{I_1} - \sum_{k}^{N_f}\left(F_{c_1}^{k} \times T_{c_1,p_1}^{I_1,k}\right)\right) \geq 0 \qquad (8.17)$$

Now, the cores c_1 and c_2 cannot be tested in parallel by taking the chunks from their corresponding set of pools. More precisely, the cores c_1 and c_2 are tested by a_1 and a_2 chunks which belong to the pools p_1 and p_2, respectively. Those chunks are transported through the IO pairs I_1 and I_2. Therefore, the chunks belonging to different cores cannot be transported in parallel by the different IO pairs because of resource conflict. Expression (8.18) has been taken care of such condition.

$$b_{c_1,p_1,a_1}^{I_1} \times b_{c_2,p_2,a_2}^{I_2}\left[m\left(S_{c_1,p_1,a_1,x_1}^{I_1} - S_{c_2,p_2,a_2,x_2}^{I_2} - \sum_{k}^{N_f}\left(F_{c_2}^{k} \times T_{c_2,p_2}^{I_2,k}\right)\right)\right.$$
$$\left. + n\left(S_{c_2,p_2,a_2,x_2}^{I_2} - S_{c_1,p_1,a_1,x_1}^{I_1} - \sum_{k}^{N_f}\left(F_{c_1}^{k} \times T_{c_1,p_1}^{I_1,k}\right)\right)\right] \geq 0 \qquad (8.18)$$

To linearize expression (8.18) unique binary variable r has been introduced to replace the term $b_{c_1,p_1,a_1}^{I_1} \times b_{c_2,p_2,a_2}$ and four additional constraints have been added, similar to (8.5). The corresponding expression is as follows:

$$r\left[m\left(S_{c_1,p_1,a_1,x_1}^{I_1} - S_{c_2,p_2,a_2,x_2}^{I_2} - \sum_{k}^{N_f}\left(F_{c_2}^{k} \times T_{c_2,p_2}^{I_2,k}\right)\right)\right.$$
$$\left. + n\left(S_{c_2,p_2,a_2,x_2}^{I_2} - S_{c_1,p_1,a_1,x_1}^{I_1} - \sum_{k}^{N_f}\left(F_{c_1}^{k} \times T_{c_1,p_1}^{I_1,k}\right)\right)\right] \geq 0 \qquad (8.19)$$

Expression (8.19) is still nonlinear. To linearize such expression, two unique binary variables u and v have been introduced to replace the terms $r \times m$ and $r \times n$ and eight additional expressions have been added, similar to (8.5).

Now, the transformed expression can be written as follows:

$$
u\left(S^{I_1}_{c_1,p_1,a_1,x_1} - S^{I_2}_{c_2,p_2,a_2,x_2} - \sum_{k}^{N_f}\left(F^k_{c_2} \times T^{I_2,k}_{c_2,p_2}\right)\right)+
$$
$$
v\left(S^{I_2}_{c_2,p_2,a_2,x_2} - S^{I_1}_{c_1,p_1,a_1,x_1} - \sum_{k}^{N_f}\left(F^k_{c_1} \times T^{I_1,k}_{c_1,p_1}\right)\right) \geq 0 \tag{8.20}
$$

To linearize expression (8.20), we have introduced four unique binary variables and four more variables. Four new equations, similar to (8.5) and sixteen new equations, similar to (8.6)–(8.7) are added.

Constraints for Sequentially Sending the Chunks of a Core During testing, all chunks of a core are transported through the single IO pair. Thus, two chunks of a core cannot be scheduled at the same time. Therefore, different chunks of a core should be scheduled at different instants of time. That is, testtime of two chunks do not overlap if and only if either one of the following conditions hold.

$$
S^I_{c,p_1,a_1,x_1} \geq S^I_{c,p_2,a_2,x_2} + \sum_{k}^{N_f}\left(F^k_c \times T^{I,k}_{c,p_2}\right) \text{ (or)}
$$
$$
S^I_{c,p_2,a_2,x_2} \geq S^I_{c,p_1,a_1,x_1} + \sum_{k}^{N_f}\left(F^k_c \times T^{I,k}_{c,p_1}\right) \tag{8.21}
$$

To represent such constraint in ILP, the binary variables g and h have been introduced, such that, $g + h = 1$. The expression (8.21) can be represented as in expression (8.22).

$$
g\left(S^I_{c,p_1,a_1,x_1} - S^I_{c,p_2,a_2,x_2} - \sum_{k}^{N_f}\left(F^k_c \times T^{I,k}_{c,p_2}\right)\right)+
$$
$$
h\left(S^I_{c,p_2,a_2,x_2} - S^I_{c,p_1,a_1,x_1} - \sum_{k}^{N_f}\left(F^k_c \times T^{I,k}_{c,p_1}\right)\right) \geq 0 \tag{8.22}
$$

$$b^I_{c,p_1,a_1} \times b^I_{c,p_2,a_2} \left[g \left(S^I_{c,p_1,a_1,x_1} - S^I_{c,p_2,a_2,x_2} - \sum_{k}^{N_f} \left(F^k_c \times T^{I,k}_{c,p_2} \right) \right) \right.$$

$$\left. + h \left(S^I_{c,p_2,a_2,x_2} - S^I_{c,p_1,a_1,x_1} - \sum_{k}^{N_f} (F^k_c \times T^{I,k}_{c,p_1}) \right) \right] \geq 0 \tag{8.23}$$

The packets of a core c can belong to different pools, such as, p_1 and p_2. Furthermore, all packets of a core can be transported by the same IO pair, I and tested at kth frequency. Such constraint can be assured by the expression (8.23). The expression (8.23) is not linear. To linearize, we have replaced $b^I_{c,p_1,a_1} \times b^I_{c,p_2,a_2}$ by the binary variable j and added four equations, similar to (8.5). Then the modified expression can be expressed as follows:

$$j \left[g \left(S^I_{c,p_1,a_1,x_1} - S^I_{c,p_2,a_2,x_2} - \sum_{k}^{N_f} \left(F^k_c \times T^{I,k}_{c,p_2} \right) \right) + \right.$$

$$\left. h \left(S^I_{c,p_2,a_2,x_2} - S^I_{c,p_1,a_1,x_1} - \sum_{k}^{N_f} \left(F^k_c \times T^{I,k}_{c,p_1} \right) \right) \right] \geq 0 \tag{8.24}$$

To linearize the expression (8.24), we have replaced $j \times g$ and $j \times h$ by the binary variables q and y and added four additional constraints, similar to (8.5). Now, the transformed expression would be as follows:

$$q \left(S^I_{c,p_1,a_1,x_1} - S^I_{c,p_2,a_2,x_2} - \sum_{k}^{N_f} \left(F^k_c \times T^{I,k}_{c,p_2} \right) \right) +$$

$$y \left(S^I_{c,p_2,a_2,x_2} - S^I_{c,p_1,a_1,x_1} - \sum_{k}^{N_f} \left(F^k_c \times T^{I,k}_{c,p_1} \right) \right) \geq 0 \tag{8.25}$$

To linearize equation (8.25), we have introduced two unique binary variables and four more variables. Four new equations, similar to (8.5) and twenty new equations, similar to (8.6)–(8.7).

The tool CPLEX Cplex (2013) has been used to solve the formulated ILP and get the optimum solution. However, excepting for very small benchmarks, it takes huge amount of CPU time to arrive at the solution. Furthermore, the inclusion of thermal constraints in the ILP is not straight forward and is quite expensive. Hence, in the following, a Particle Swarm Optimization (PSO) based technique with its variants have been proposed to find the solutions for bigger benchmarks, producing results within a reasonable amount of time, including thermal-aware test schedule.

4 PSO Formulation for Preemptive Test Scheduling

This section presents a Particle Swarm Optimization (PSO) based solution strategy for the multi-frequency based preemptive test scheduling problem. The particle formulation of the problem and its fitness calculation have been detailed next.

4.1 Particle Structure and Fitness

A particle represents a candidate test schedule of cores. It has four components: core part, I/O part, frequency part and preemption part. An example of a particle has been shown in Fig. 8.3. The core part corresponds to a permutation of cores C_1 to C_n. It represents an order in which scheduler will pick up the cores for probable scheduling and assign it to the corresponding time slot. The next part is an array of I/O pairs. It is of size same as the core part. If the k number of I/O pairs are available, each entry contains an integer between 1 and k. The third part contains an array of size equal to the number of cores. If m frequencies are available, each entry of this array contains one of the available m frequencies. Each preemption part entry is a number between 0 and 1. For non-preemptive cores, the entry is always 1. For a preemptive core, if the corresponding entry is x and the core has a total of p test patterns, $x * p$ number of patterns will be tested as a single block. Thus, testing of the entire core is distributed over a number of blocks, each block being scheduled independently of others. However, since the same I/O pair is attached to the core in each test session, the same routing resources will be utilized for them. To compute the fitness of a particle, the scheduler starts with the first core from the core part. At time instant zero the core is scheduled to start its test. Throughout a core test, proper NoC resources, such as links and routers remain reserved and during preemption reserved resource can be released. The proper resources are decided by the I/O pair, core-under-test and routing algorithm. At a certain time instant, suppose that the scheduler has scheduled up to the ith core in the core part of a particle. Now, scheduling time of $(i + 1)$th core will be decided by consulting the corresponding I/O pair and the frequency at which it will be tested, that is, $(i + 1)$th entries in the I/O pair and frequency parts. If the corresponding I/O pair is k, the next available time slot for k is determined by the block $(x \times p)$, where x represents the preemption coefficient in the $(i + 1)$th core and p denotes total test packets for that core, so that proper resources are available for a test of $(i+1)$th core. The scheduler can schedule

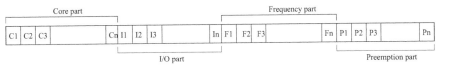

Fig. 8.3 Particle structure for test scheduling

the cores until there is no test packet left. The overall testtime of the entire schedule is computed from the highest time for any of the I/O pairs. The testtime of each chunk is computed by the expression 8.1.

4.2   Evolution Process

Particles evolve through generations to generate new particles. It is expected that the new particle will produce a better solution. In the first generation, each particle is created randomly and local best ($lbest^k$) of each particle is set to itself. The global best ($gbest_i$) for a generation is elected as a particle which has the minimum fitness value from the set of local bests.

Based on the evolution policy, PSO can be classified into two groups: discrete PSO (DPSO) and continuous PSO. The continuous PSO evolves particle positions by updating their velocities in each dimension. In the particle, different parts are evolved either following the strategy discrete or continuous PSO. The core, I/O pair and frequency parts of the particle are evolved using the strategy followed in discrete PSO which has been discussed in Chap. 5. However, the preemption part is evolved using the strategy followed in continuous PSO using the expression (7.5). Every particle evolves and adjusts its local best accordingly. The global best is also calculated from the set of local best values.

4.3   Augmentation to the Basic PSO

To improve quality of the solution produced by the basic PSO, several augmentations, such as use of an efficient random number generator, perform inversion mutation (IM) operation to each particle and multi-PSO have been incorporated into it. Such augmentations have already been discussed in Chap. 5. The maximum number of the PSO runs to be executed has been set as follows:

- A user-defined value for the maximum number of PSO runs. In this experiment, it has been taken as 20 PSO runs.
- The global best fitness does not change in the last 100 PSO runs.

The complete PSO engine has been described in Algorithm 8.1. In this algorithm, each *Particle* represents a configuration corresponding to scheduling order of cores, along with the corresponding I/O pair, frequency and preemption coefficient. The algorithm generates *NPart* number of random particles in the initial generation of PSO (line 10). The local best of individual particles is set to the particle itself. The global best particle is set to be the best configuration found in the local best configuration set (line 21). Each configuration is evaluated using the expression (8.1). Here, *MGEN* is the maximum number of generations for which individual PSOs may run. The total number of PSO runs is represented by *MPSO*.

Algorithm 8.1 PSO-based test scheduling approach for NoC-based systems

Input: Topology graph T

Output: Test scheduling order

1: Set $MGEN$, $MPSO$ and $NPart$
2: Set W
3: $BestFitness \leftarrow \infty$
4: **for** m from 0 to $MPSO$ **do**
5:     $IM \leftarrow true$
6:     $BeforeIMBestFitness \leftarrow \infty$
7:     **while** $IM = true$ **do**
8:         $gen \leftarrow 0$
9:         **while** $gen < MGEN$ and $IM = true$ **do**
10:             **for** p from 0 to $NPart$ **do**
11:                 **if** $gen = 0$ and $m = 0$ **then**//Initialization
12:                     $Particle_p \leftarrow \{Random()\}$
13:                     $Particle_p^{pbest} \leftarrow Particle_p$
14:                 **else**//Update particle
15:                     $UpdatePart(Particle_p, Particle_p^{pbest}, Particle^{gbest}, ProbRandom_{pbest}, ProbRandom_{gbest})$
16:                     Compute fitness of $Particle_p$ using expression (8.1)
17:                     $UpdateLBest(Particle_p^{pbest}, Particle_p)$
18:                 **end if**
19:             **end for**
20:         Compute global best particle $Particle^{gbest}$ from the set of particles $Particle^{pbest}$
21:         Update the $BestFitness$ as global best as per the requirement and reset the gen
22:         counter
23:         **if** $BeforeIMBestFitness = BestFitness$ **then**
24:             $IM \leftarrow false$
25:         **end if**
26:         $gen \leftarrow gen + 1$
27:         **end while**

28:     **if** $IM = true$ **then**
29:         $BeforeIMBestFitness \leftarrow BestFitness$
30:         Perform Inverse Mutation (IM) at random position of each particle of the last generation
31:     **end if**
32:     **end while**
33:     **for** p from 0 to $NPart$ **do**
34:         $Particle_p \leftarrow \{Random()\}$
35:         $CopyLbest(Particle_p^{pbest})$
36:     **end for**
37: **end for**

After generating the initial configurations and finding the global best configuration at first generation, the particles are evolved using *UpdatePart( )*. Each particle evolves by sharing the experience of its local as well as the global best of the generation with some random probabilities. The *swap* sequences are generated for core, I/O pair and frequency parts of the particle, local and global best configurations (as discussed in Chap. 5). The preemption part of the schedule gets updated based on the particle, local and global best particles. A new particle or configuration is generated from the existing one by applying the swap sequences subject to some probability for core, I/O and frequency part. However, the preemption part of a particle gets modified to values closer to the best particles, with some probability. Furthermore, the local best of each particle gets changed by comparing the fitness between its current *lbest* and the new particle (line 17). After each generation, the PSO engine checks the *BestFitness* value. The current generation is reset to the initial generation, if the PSO gets a better solution than the earlier one. Otherwise, it keeps a count of the generations. The *inversion mutation* (IM) operator (line 29) is applied to each particle, if generation count reaches *MGEN* and *IM* is *true*. For multiple PSO runs, the PSO engine creates the new particles by assigning random configurations, whereas, local best for the particles are passed from the previous PSO run. The PSO engine stops when multiple PSO count reaches the maximum value, (*MPSO*).

5   PSO for Thermal-Aware Preemptive Test Scheduling

The inclusion of thermal constraints in the ILP (presented in Sect. 3) is not straight forward and is quite expensive. In order to get near-optimal solutions for large benchmarks, the proposed PSO has been extended for generating thermal-aware test schedules. The particle structure is similar to the previous one and the particles evolved in a similar fashion. However, the fitness of individual particles have been measured by the expression (8.26)

$$Fitness = W \times \frac{T_{peak}}{\alpha} + (1 - W) \times \frac{T}{\beta} \qquad (8.26)$$

where, T_{peak} is the peak temperature and T is the overall testtime of the entire schedule. These two quantities are evaluated using the expressions (8.2) and (8.1). W is a weighting coefficient meant to balance the optimization of testtime and peak temperature of the die. More precisely $W \epsilon [0, 1]$. When $W = 1$, it minimizes the temperature of the die and for $W = 0$, it considers only testtime to minimize. As testtime and temperature have different units, normalization of these two metric has been done by assuming worst-case scenarios. To set the value of α, we perform test scheduling of all cores, assuming the availability of only one I/O pair. For each I/O pair, the evaluation is carried out. The maximum among these gives the normalization factor α. β is fixed to peak temperature reached, assuming that all cores are being tested in parallel.

6 Experimental Results

In this section, we have presented the results of our experimentation with ITC'02 SoC benchmarks Marinissen et al. (2002), d695, p22810, p93791 and a synthetic benchmark for 64 cores. The synthetic benchmark, syn-64 has been generated from the core repository. The core repository contains cores taken from ITC'02 benchmarks system. Thus, all the test information of the cores are preserved. However, cores are chosen randomly to create such synthetic benchmarks and cores are mapped onto the mesh-based 2D and 3D multi-core systems. The dimension of each benchmark has been presented in Table 8.2. For each of them, we have generated the floorplan manually. The detailed experimental setup has been presented below.

6.1 Experimental Setup

Router and link power values have been calculated using tool *DSENT* Sun et al. (2012) for 45-nm technology node. In this experiment, core powers have been generated using the distribution shown in Table 8.2. Leakage power (assumed to be 10% of active power Segars (1997)) consumptions for core, router and link have also been taken into consideration. Furthermore, individual tiles have dimension $1 \times 1\,mm^2$ similar to Manna et al. (2018). The heat sink is connected via heat spreader to the bottom layer for 3D and has thickness of $200\,\mu m$ and other active (power dissipating) layers are assumed to be thinned to $50\,\mu m$ for better heat conductivity Cheng et al. (2013). A $10\,\mu m$ thin layer containing thermal interface material (TIM) is used in between two layers. For vertical communication, each TSV bundle contains 8×9 TSVs (32 data lines, 1 request, 1 grant, 1 read and 1 write clock lines and for dual connection). In our considered core size ($1 \times 1\,mm^2$), $0.3 \times 0.3\,mm^2$ is allocated for the router and rest of the area for processor and

Table 8.2 Power and mesh-size information of benchmarks

Benchmarks	Power information (mW)			No. of cores	2D/3D Mesh dimension
	Max.	Min.	Mean		
d695	3139	255	1732.53	10	4×3
p22810	6213	176	3698.12	28	4×7
p93791	7015	554	3123.56	32	$8 \times 4, 4 \times 4 \times 2, 4 \times 2 \times 4$
syn-64	6938	554	3012.58	64	$8 \times 8, 4 \times 8 \times 2, 4 \times 4 \times 4$

Table 8.3 2D and 3D IC parameters

Parameters	Value
Technology node [nano-m]	45
Tile Size [mm × mm]	1 × 1
TSV diameter [μm]	10
TSV pitch [μm]	20
Bottom layer thickness [μm]	200
Non-bottom layer thickness [μm]	50
TIM layer thickness [μm]	10
Heat sink side/thickness [mm]	14 × 14 × 10

Table 8.4 Thermal simulation parameters

Parameters	Value
Silicon thermal conductance [W/(m K)]	150
Silicon specific heat [J/(m^3 K)]	1.75×10^6
TIM thermal capacitance [W/(m K)]	4
TIM specific heat [J/(m^3 K)]	4×10^6
TSV thermal conduction [W/(m K)]	300
TSV Specific heat [J/(m^3 K)]	3.5×10^6
Convection resistance to ambient [K/W]	3.0
Heat sink thermal conduction [W/(m K)]	400
Heat sink specific heat [J/(m^3 K)]	3.55×10^6
Ambient temperature [C]	45

memory. The area allocated for router (0.3×0.3 mm^2) is used to place the 8×9 TVSs and is sufficient to accommodate them. Some other important physical properties of the 3D IC model are summarized in Table 8.3.

The tool *HotSpot* Huang et al. (2006) is used to get the thermal resistance for 2D as well as 3D multi-core systems and corresponding peak temperature is calculated following the procedure mentioned in Sect. 1. The used thermal simulation parameters are presented in Table 8.4.

6.2 Testtime Comparison Between ILP and PSO

To check the quality of the proposed PSO approach for the preemptive test scheduling problem, we first compare the ILP and PSO results and execution time. For synthetic benchmarks S1 (4 cores and test packets {3, 2, 2, 2}), S2 (4 cores and test packets {4, 4, 3, 3}), S3 (4 cores and test packets {5, 4, 5, 5}) and S4 (6 cores and test packets {3, 2, 2, 3, 2, 2}), PSO could obtain the same solution such as 154 (for S1), 201 (for S2), 277 (for S3) and 552 (for S4), reported by ILP, with less time as compared to ILP. In this experiment two IO pairs have been used for test delivery. The synthetic benchmarks are mapped onto 2D mesh-based-NoC multi-core systems. For other ITC'02 benchmarks, ILP could not start or complete due to the creation of large number of constraints. The CPLEX Cplex (2013) tool has been used to solve the formulated ILP. Both ILP and PSO have been implemented on a Dell PowerEdge T410 system with 8 cores (Intel Xeon processor, E5606@2.12 GHz), 64 GB RAM. The capacity of the PSO to achieve the optimal results, found from ILP, corroborates the quality of the PSO.

6.3 Effect of Inversion Mutation (IM) and Randomness into the Basic PSO

The effect of augmentations to the basic PSO, such as, IM and randomness of random number generator have been presented in this section. The corresponding results have been noted in Table 8.5 for the benchmarks p93791 and syn-64. In this experiment four IO pairs have used for test delivery. The second column notes the results of basic PSO without any augmentation. The third and fourth columns show the results of incorporation of inverse mutation and SIMT into this basic PSO. Table 8.5 projects better testtime while using the technique IM and the SIMT random generator with the basic PSO. The result reported here are for $W = 0$, that is, fully minimize the testtime. This establishes the effectiveness of our proposed augmentation strategies for improving the solution quality.

6.4 Impact of Multiple PSOs

The effect of running multiple PSOs on testtime reduction has been presented in Table 8.6 for the ITC'02 benchmark circuits p22810 and p93791. In this experiment four IO pairs have been used, marked as 4 under the 'IO pairs' column. From Table 8.6, it can be seen that multiple PSO performs better than the basic one.

Table 8.5 Testtime comparison of different augmentation techniques

Benchmarks	Basic PSO	Basic PSO with IM only	Basic PSO with SIMT only
p93791	112,295	98,169	97,316
syn-64	213,631	187,654	186,645

Table 8.6 Testtime comparison between basic and multiple PSO

Benchmarks	IO pairs	Testtime (clock cycles)	
		Basic PSO	Multiple PSO
p93791	4	112,295	85,967
syn-64	4	213,631	150,123

Table 8.7 Preemptive vs. non-preemptive test scheduling technique

Benchmarks	IO pairs	Proposed multiple PSO-based approaches	
		Preemptive	Non-preemptive
d695	2	9807	9807
	3	6933	6933
	4	5272	5272
p22810	2	145,098	145,563
	3	98,167	104,462
	4	85,453	97,644
p93791	2	159,188	160,773
	3	110,338	121,089
	4	85,967	114,317
syn-64	2	251,376	281,453
	3	182,485	223,241
	4	150,123	183,245

6.5 Testtime Comparison with Non-Preemptive Method

To compare the testtime between proposed preemptive and the non-preemptive test scheduling strategy, the authors have implemented the non-preemptive version as well. The corresponding testtime values have been noted in Table 8.7, when $W = 0$. The ITC'02 benchmarks are used for this experiment. For three different IO pairs, 2, 3 and 4, authors have reported the corresponding testtime values for the benchmarks. It is observed that testtime decreases with the increase in number of IO pairs. Further, it can be noted that the proposed preemptive approach reduces testtime 11.25% better as compared to that of non-preemptive one.

6.6 Trading-Off Between Peak Temperature and Testtime

Figures 8.4, 8.5 and 8.6 show the behavior of the peak temperature and testtime during test of benchmarks: d695, p22810 and p93791. For benchmark p22810, the power budget has set at 20 Watt. We have varied the parameter W to generate a set of solutions. The value $W = 0$ is expected to produce results optimized towards testtime minimization, while $W = 1$ is expected to perform temperature optimization. For three different number of IO pairs, 2, 3 and 4, we have reported the corresponding testtime values of the SoC and the peak temperature attained. It can be noticed that $W = 0$ has resulted in the minimum testtime. As W increases, testtime degrades, while the peak temperature reduces. Similar trend can be found for other benchmarks, such as d695 and p93791 reported in Fig. 8.4 and 8.6, for power budget 11 Watt and 20 Watt.

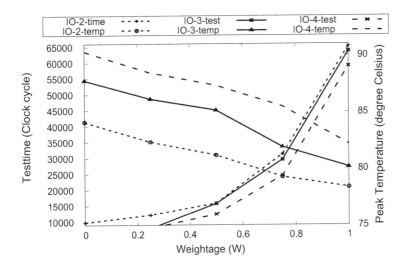

Fig. 8.4 Temperature vs. testtime for d695

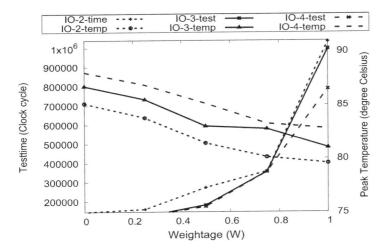

Fig. 8.5 Temperature vs. testtime for p22810

6.7 Comparison with Other Methods

Table 8.8 shows the testtime comparison of proposed approach with the works
Liu and Iyengar (2006), Liu et al. (2006) and Manna et al. (2015). We have also
compared the peak temperature with the work reported in Manna et al. (2015).
Works Liu and Iyengar (2006) and Liu et al. (2006) take the peak temperature
as a constraint and produce a test schedule that honor the temperature limit. To
compare between the testtimes, we have fed the peak temperature value reported by
our proposed method in a specific case to Liu and Iyengar (2006) and Liu et al.
(2006) as the temperature constraint. The testtime reported by Liu and Iyengar

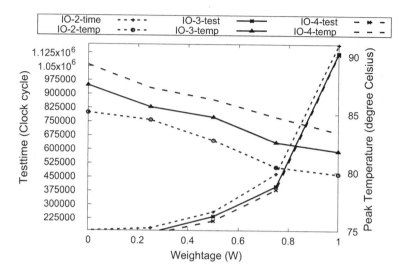

Fig. 8.6 Temperature vs. testtime for p93791

(2006) and Liu et al. (2006) are then compared with the testtime of our approach.
For example, the benchmark p22810 with $W = 0.5$ and IO pairs 3, the peak
temperature reported by our formulation is $83.01°$ C. This has associated testtime
of 182,060. On the other hand, the works Liu and Iyengar (2006) and Liu et al.
(2006) when fed with a peak temperature constraint of $83.01°$ C, generate a schedule
with testtime 306,445 and 264,765, respectively. In Manna et al. (2015), a model
has been used to compute the temperature. Thus, it provides high temperature and
testtime compared to the proposed method. Taking the results for proposed method
as unity, other approaches, like Manna et al. (2015), Liu et al. (2006) and Liu and
Iyengar (2006) take 19%, 55% and 61% more testtime, on an average, to test the
system and the method Manna et al. (2015) produces 32% more peak temperature
for $W = 0$. Similarly, for $W = 0.5$, other methods, such as Manna et al. (2015),
Liu et al. (2006) and Liu and Iyengar (2006) take 17%, 48% and 59% more testtime
and work Manna et al. (2015) produce 37% more temperature. In Table 8.8, unity
in the rank field indicates the best testtime and peak temperature achievable by our
proposed method.

6.8   Thermal-Aware Test Scheduling for 3D NoC-Based Systems

In this experiment, we have considered symmetrically partially connected 3D mesh
multi-core systems Liu et al. (2011a) with 2 and 4 layers. That is, each router in the
layer has no vertical connection. The experimental results are presented below.

Table 8.8 Comparison with test strategies Manna et al. (2015), Liu et al. (2006) and Liu and Iyengar (2006)

Benchmarks	IO pair	PSO ($W = 0$)				PSO ($W = 0.5$)			
		Proposed	$Method_1$	$Method_2$	$Method_3$	Proposed	$Method_1$	$Method_2$	$Method_3$
		$\mathbb{T}$, T_{peak}	$\mathbb{T}$, T_{peak}	$\mathbb{T}$	$\mathbb{T}$	$\mathbb{T}$, T_{peak}	$\mathbb{T}$, T_{peak}	$\mathbb{T}$	$\mathbb{T}$
d695	2	9807, 84.1	9807, 106.23	14,315	13,225	15,859, 81.16	16,859, 107.20	18,345	19,530
	3	6933, 87.76	6933, 109.27	10,708	7955	15,771, 85.14	17,139, 110.10	22,567	20,614
	4	5272, 90.31	5272, 113.23	9578	7121	12,405, 87.32	14,312, 114.10	16,291	14,620
p22810	2	145,028, 85.12	195,565, 115.30	202,889	240,309	273,510, 81.42	303,715, 115.10	376,298	967,732
	3	98,167, 86.73	124,460, 119.50	153,530	173,852	182,060, 83.01	251,145, 119.10	306,445	264,765
	4	85,453, 88.01	97,648, 122.20	115,081	154,329	174,499, 85.12	213,120, 126.60	280,345	253,010
p93791	2	159,188, 85.14	200,775, 112.30	229,649	283,121	260,852, 82.74	313,215, 112.20	487,989	453,642
	3	110,338, 87.47	151,092, 115.40	172,960	189,740	235,955, 84.74	262,710, 115.20	354,095	281,710
	4	85,967, 89.25	114,319, 119.00	155,189	164,638	211,580, 86.24	248,215, 119.20	285,470	255,449
Rank		**1.00, 1.00**	**1.19, 1.32**	**1.55**	**1.61**	**1.00, 1.00**	**1.17, 1.37**	**1.48**	**1.59**

$\mathbb{T}$ Testtime (Clock Cycles), T_{peak} Peak temperature (°C), $Method_1$ Manna et al. (2015), $Method_2$ Liu et al. (2006), $Method_3$ Liu and Iyengar (2006)
Bold values indicate the entire results presented in this table

6.8.1 Trading-Off Peak Temperature and Testtime

Table 8.9 shows the trade-off results for benchmarks. We have varied the weight parameter, W, to generate solutions. Three IO pairs are noted in the table as 2, 3 and 4. The value $W = 0$ is expected to produce results optimized towards testtime minimization, while $W = 1$ is expected to perform thermal optimization. It can be noticed that $W = 0$ has resulted in the minimum testtime. As W increases, testtime degrades, while the peak temperature improves. The testtime decreases as IO pairs increase.

6.8.2 Comparison with Other Methods

Finally, Table 8.10 shows the testtime comparison of proposed approach with the works Liu and Iyengar (2006), Liu et al. (2006) and Manna et al. (2015) also shows the peak temperature and testtime comparison. The works, Liu and Iyengar (2006), Liu et al. (2006) and Manna et al. (2015) have been originally proposed to test the 2D multi-core systems. For comparison purpose, we have extended those works for 3D environment. Works Liu and Iyengar (2006) and Liu et al. (2006) take the peak temperature as a constraint and generate a test schedule that does not violate it. To compare between the testtimes, we have fed the peak temperature value reported by our tool in a specific case to Liu and Iyengar (2006) and Liu et al. (2006) as the temperature constraint. The testtime reported by Liu and Iyengar (2006) and Liu et al. (2006) are then compared with the testtime of our approach. For example, benchmark p93791 with $W = 0.5$, IO pairs 4 and mapped onto 3D NoC having two layers, the peak temperature reported by our formulation is $82.41° \text{C}$. This has associated testtime of 169,296. On the other hand, the works Liu and Iyengar (2006) and Liu et al. (2006) when fed with a peak temperature constraint of $82.41° \text{C}$, generate a schedule with testtime 209,814 and 208,145, respectively. The proposed strategy reduces temperature as well as testtime compared to the work in Manna et al. (2015). The work, in Manna et al. (2015), used a model to calculate the temperature. Taking the results for proposed method as unity, other approaches, like Manna et al. (2015), Liu et al. (2006) and Liu and Iyengar (2006) take 23%, 49% and 62% more testtime, on an average, to test the system and the method Manna et al. (2015) produces 22% more peak temperature for 3D multi-core systems with 2 layers and $W = 0$. Similarly, for 3D NoC system with 2 layers and $W = 0.5$, other methods, such as Manna et al. (2015), Liu et al. (2006) and Liu and Iyengar (2006) take 13%, 43% and 45% more testtime and work Manna et al. (2015) produce 27% more temperature. Similar trend can be seen when considered 3D multi-core systems with four layers. In Table 8.10, unity in the rank field indicates the best testtime and peak temperature achievable by our proposed method.

Table 8.9 Trade-off system safety and testtime

Layers	Two						Four					
IO pairs	2		3		4		2		3		4	
W	$\mathbb{T}, T_{peak}$		$\mathbb{T}, T_{peak}$		$\mathbb{T}, T_{peak}$		$\mathbb{T}, T_{peak}$		$\mathbb{T}, T_{peak}$		$\mathbb{T}, T_{peak}$	
Benchmark: p93791 (for power budget 21 Watt)												
0	132,134, 83.21		106,156, 84.15		78,698, 86.45		112,345, 88.54		79,421, 89.56		68,125, 90.24	
0.2	163,257, 82.73		127,625, 83.15		98,632, 84.12		126,789, 85.32		99,362, 86.18		93,199, 87.36	
0.5	246,890, 80.15		220,518, 81.74		169,296, 82.41		198,931, 83.98		168,630, 85.14		139,812, 86.18	
0.8	309,851, 78.35		259,312, 79.25		208,729, 81.44		259,625, 82.01		207,965, 84.32		188,097, 85.73	
1	673,210, 76.92		485k116, 77.14		398,973, 78.11		619,316, 80.35		416,772, 81.84		378,925, 82.52	
Benchmark: syn-64 (for power budget 24 Watt)												
0	139,156, 81.62		100,125, 83.43		93,862, 85.98		161,365, 87.24		112,168, 88.46		89,761, 90.01	
0.2	162,310, 79.23		119,560, 81.54		127,531, 82.64		182,168, 85.75		148,967, 86.34		138,725, 87.34	
0.5	278,362, 77.59		178,675, 80.12		168,329, 81.16		259,601, 83.34		227,390, 84.65		219,216, 85.32	
0.8	359,167, 76.24		359,781, 78.27		349,705, 79.61		459,865, 80.61		398,602, 82.33		329,650, 83.22	
1	1,039,965, 74.35		9,989,752, 75.23		789,642, 77.53		1,167,361, 79.11		1,118,275, 81.35		1,108,652, 82.86	

$\mathbb{T}$ Testtime (Clock Cycles), T_{peak} Peak temperature (°C)

Table 8.10 Comparison with test strategies Manna et al. (2015), Liu et al. (2006) and Liu and Iyengar (2006)

Layers	Benchmarks	IO pairs	PSO (w=0)				PSO (w=0.5)			
			Proposed $\mathbb{T}$, T_{peak}	$Method_1$	$Method_2$ $\mathbb{T}$	$Method_3$	Proposed $\mathbb{T}$, T_{peak}	$Method_1$	$Method_2$ $\mathbb{T}$	$Method_3$
Two	p93791	2	132,134, 83.21	183,215, 93.65	208,250	218,502	249,890, 80.15	268,319, 93.65	299,812	338,642
		3	106,156, 84.15	148,625, 100.32	159,820	168,214	220,518, 81.74	249,912, 103.17	309,732	288,421
		4	78,698, 86.45	81,315, 105.61	109,212	120,324	169,296, 82.41	188,215, 109.23	209,814	208,145
	Syn-64	2	139,156, 81.62	179,260, 99.25	209,756	219,723	278,362, 77.59	287,934, 95.13	389,723	427,423
		3	100,125, 83.43	108,251, 107.47	158,623	177,324	178,675, 80.12	219,345, 105.62	279,423	321,865
		4	93,862, 85.98	109,621, 111.31	128,913	148,346	168,329, 81.16	198,315, 109.34	298,971	247,453
Rank			**1.00, 1.00**	**1.23, 1.22**	**1.49**	**1.62**	**1.00, 1.00**	**1.13, 1.27**	**1.43**	**1.45**
Four	p93791	2	112,345, 88.54	137,512, 95.32	177,912	197,615	198,931, 83.98	238,231, 95.12	318,345	357,432
		3	79,421, 89.56	89,873, 104.45	108,123	99,812	168,630, 85.14	217,123, 106.34	287,346	330,123
		4	68,125, 90.24	70,312, 109.32	99,314	89,627	139,812, 86.18	168,134, 111.64	219,431	197,321
	Syn-64	2	161,365, 87.24	178,614, 101.16	272,346	288,837	259,601, 83.34	308,645, 98.35	382,461	425,418
		3	112,168, 88.46	138,945, 109.16	189,321	204,127	227,390, 84.65	268,743, 107.65	327,345	358,421
		4	89,761, 90.01	97,614, 114.11	139,932	147,251	219,216, 85.32	237,612, 113.48	279,931	259,964
Rank			**1.00, 1.00**	**1.14, 1.19**	**1.56**	**1.60**	**1.00, 1.00**	**1.19, 1.24**	**1.51**	**1.59**

$\mathbb{T}$ Testtime (Clock Cycles), T_{peak} Peak temperature (°C), $Method_1$ Manna et al. (2015), $Method_2$ Liu et al. (2006), $Method_3$ Liu and Iyengar (2006)
Bold values indicate the entire results presented in this table

7 Conclusion

In this chapter, we have presented preemptive and non-preemptive schedule strategies to test the cores in 2D and 3D NoC-based systems. A core can support several test frequencies. It has been shown that the preemptive test strategy has the potential to reduce the testtime, compared to the non-preemptive version. An ILP has been formulated for the preemptive testing strategy. It has been seen that ILP is suitable for small benchmarks. Next, a PSO-based strategy has been proposed to solve the problem for larger NoCs. The PSO has been also extended for thermal-aware testing to improve the yield. The experimental results show improvement in terms of testtime reduction compared to other strategies reported in literature.

References

Cheng, Y., Zhang, L., Han, Y., & Li, X. (2013). Thermal-constrained task allocation for interconnect energy reduction in 3-D homogeneous MPSoCs. *IEEE Transactions on Very Large Scale Integration (VLSI) Systems, 21*(2), 239–249.

Cplex (2013). www.ibm.com/software/in/integration/optimization/cplex.

Dubois, F., Sheibanyrad, A., Petrot, F., & Bahmani, M. (2013). Elevator-First: A deadlock-free distributed routing algorithm for vertically partially connected 3D-NoCs. *IEEE Trans on Computers, 62*(3), 609–615.

Huang, W., Ghosh, S., Velusamy, S., Sankaranarayanan, K., Skadron, K., & Stan, M. (2006). HotSpot: A compact thermal modeling methodology for early-stage VLSI design. *IEEE Trans on Very Large Scale Integration (VLSI) Systems, 14*(5), 501–513.

Iyengar, V., Chakrabarty, K., & Marinissen, E. J. (2002). Test wrapper and test access mechanism co-optimization for system-on-chip. *Journal of Electronic Testing, 18*(2), 213–230.

Kiamehr, S., Tahoori, M. B., & Anghel, L. (2018). *Manufacturing threats*. Springer, Berlin.

Liu, C., & Iyengar, V. (2006). Test scheduling with thermal optimization for network-on-chip systems using variable-rate on-chip clocking. In *Proceedings of the Conference on Design, Automation and Test in Europe (DATE)* (pp. 6–10).

Liu, C., Iyengar, V., & Pradhan, D. K. (2006). Thermal-aware testing of network-on-chip using multiple-frequency clocking. In *Proceedings of VLSI Test Symposium* (pp. 46–51).

Liu, C., Zhang, L., Han, Y., & Li, X. (2011a) Vertical interconnects squeezing in symmetric 3D mesh network-on-chip. In *Proceedings of the 16th Asia and South Pacific Design Automation Conference (ASP-DAC)* (pp. 357–362).

Manna, K., Mukherjee, P., Chattopadhyay, S., & Sengupta, I. (2018). Thermal-aware application mapping strategy for network-on-chip based system design. *IEEE Transactions on Computers, 67*(4), 528–542.

Manna, K., Reddy, C., Chattopadhyay, S., & Sengupta, I. (2015). Thermal-aware multifrequency network-on-chip testing using particle swarm optimisation. *International Journal of High Performance Systems Architecture (IJHPSA), 5*(3), 141–152.

Marinissen, E., Iyengar, V., & Chakrabarty, K. (2002). A set of benchmarks for modular testing of socs. In *Proceedings. International Test Conference (ITC)* (pp. 519–528).

Segars, S. (1997). ARM7TDMI power consumption. *IEEE Micro, 17*(4), 12–19.

Sun, C., Chen, C. O., Kurian, G., Wei, L., Miller, J., Agarwal, A., et al. (2012). DSENT - A tool connecting emerging photonics with electronics for opto-electronic networks-on-chip modeling. In *2012 IEEE/ACM Sixth International Symposium on Networks-on-Chip* (pp. 201–210).

Chapter 9
Conclusion and Future Works

1 Conclusion

NoC has evolved as a standard to solve the communication problem between cores in a SoC. Application mapping constitutes a very important step in a Network-on-Chip (NoC) based implementation of an application. For 3D NoCs, mapping should be performed together with TSV placement to ensure good performance. This thesis focuses on application mapping together with TSV placement strategy that enhances the performance of 3D NoC-based system. Several mapping strategies, integrated with TSV placement have been reported and the performance of the developed strategies has been compared. The strategies developed can broadly be classified into four categories (discussed in Chap. 2).

(a) Exact methods using Integer Linear Programming.
(b) Iterative improvement based strategies.
(c) Constructive heuristics.
(d) Meta-search techniques, such as Particle Swarm Optimization (PSO).

All the above algorithms have been proposed to minimize the communication cost. Algorithms which minimize communication cost may not consider thermal effects of that solution. To address the same, we have proposed thermal-aware mapping algorithms for 2D NoC-based system to minimize both the objectives for a given application. The peak temperature is reduced with little sacrifice in communication cost. In this direction, this dissertation has proposed solutions—an exact method and a PSO-based approach. For 3D NoC-based system, two variations of the problem have been solved—deploying the thermal vias into the given extra space in the die and without deploying the thermal vias.

Now, testing plays a key role to test such manufactured chip. Yield can be improved by testing the system quickly and correctly. Furthermore, thermal-aware test also increases the yield of the system. Thus, a thermal-aware test scheduling technique for 2D as well as 3D NoC-based systems has proposed. In this direction,

© Springer Nature Switzerland AG 2020
K. Manna, J. Mathew, *Design and Test Strategies for 2D/3D Integration for
NoC-based Multicore Architectures*, https://doi.org/10.1007/978-3-030-31310-4_9

the authors have proposed an exact method to test the 2D NoC system and a PSO-based method for 2D as well as 3D NoC-based systems.

Taking all these observations into consideration, the following conclusions can be drawn:

(a) The iterative application mapping together with TSV placement technique using KL-partitioning works well.
(b) The constructive mapping together with TSV placement technique produces encouraging solutions with less CPU time.
(c) The extended PSO technique produces the best results.
(d) Thermal-aware application mapping for 2D NoC performs better than other approaches reported in the literature.
(e) Thermal-aware application mapping and physical design for 3D NoC-based system improves the thermal safety of the system.
(f) Thermal-aware test strategies for those NoC-based systems improve the testtime as well as the system safety.

2 Possible Future Works

This book presents application mapping together with TSV placement strategies for 3D NoC to improve the performance. Thermal-aware application mapping strategies have also been proposed for 2D as well as 3D NoCs. To improve the testtime and thermal safety of the 2D as well as 3D NoC-based system, this book has proposed the preemptive test techniques. However, some important issues have not been covered in this book, which are kept as a future work. In the following we have discussed such issues:

(a) **Application-specific NoC:** The application-specific NoC design problem takes as input the chip floorplan, library of NoC components and communication requirements between the tasks of the application. It synthesizes and outputs the positions of the routers in the floorplan. Irregular and custom topologies are needed to handle the challenges in terms of irregular core sizes, different core locations, and communication flow requirements. An irregular topology may not have all links and routers in it. The techniques proposed in this thesis can be extended to custom topologies by suitable modifications, communication cost computation and candidate position selection for mapping.
(b) **Integrated mapping and scheduling:** The process of application mapping answers the question *where*, but to answer *when*, scheduling is required. If tasks of an application are mapped onto one core, task scheduling is required. Given an application task graph mapped onto NoC architecture, scheduling is the time ordering of tasks and communications determining the order in which tasks and transactions between them are to be executed such that the deadlines are met and some design parameters are optimized. This has to be explored for application mapping.

(c) **Reliability and fault tolerant technique:** The current VLSI technology is sensitive to fault/unreliablity—transient, intermittent and permanent faults. Furthermore, such technology is also sensitive to crosstalk, electro-migration interference and neutron and alpha particle strikes and power supply disturbance.

Fault tolerant methods can be used to overcome such problem at difference level of abstraction from circuit to system level. Application mapping can be used to improve the system reliability. This has to be explored in future.

(d) **Multi-cast based NoC testing:** It has been seen that IP-cores are homogenous in NoC. That is, some cores are similar in nature in NoC. Thus, those cores can be tested with similar test patterns. Such test packet can be sent in multi-cast fashion from the source which can reduce the testtime. This has to be explored in future.

Index

© Springer Nature Switzerland AG 2020

K. Manna, J. Mathew, *Design and Test Strategies for 2D/3D Integration for NoC-based Multicore Architectures*, https://doi.org/10.1007/978-3-030-31310-4